UNITED NATIONS CONFERENCE ON TRADE A

REVIEW
OF MARITIME TRANSPORT
2021

UNITED NATIONS
Geneva, 2021

© 2021, United Nations

All rights reserved worldwide

Requests to reproduce excerpts or to photocopy should be addressed to the Copyright Clearance Centre at copyright.com.

All other queries on rights and licences, including subsidiary rights, should be addressed to:

United Nations Publications
405 East 42nd Street, New York, New York
10017 United States of America
Email: publications@un.org
Website: https://shop.un.org

The designations employed and the presentation of material on any map in this work do not imply the expression of any opinion whatsoever on the part of the United Nations concerning the legal status of any country, territory, city or area or of its authorities, or concerning the delimitation of its frontiers or boundaries.

Mention of any firm or licensed process does not imply the endorsement of the United Nations.

This publication has been edited externally.

United Nations publication issued by the United Nations Conference on Trade and Development.

UNCTAD/RMT/2021

ISBN: 978-92-1-113026-3
eISBN: 978-92-1-000097-0
ISSN: 0566-7682
eISSN: 2225-3459
Sales No. E.21.II.D.21

Acknowledgements

The *Review of Maritime Transport 2021* was prepared by UNCTAD under the overall guidance of Shamika N. Sirimanne, Director of the Division on Technology and Logistics of UNCTAD, and under the coordination of Jan Hoffmann, Head of the Trade Logistics Branch, Division on Technology and Logistics. Regina Asariotis, Gonzalo Ayala, Mark Assaf, Celine Bacrot, Hassiba Benamara, Dominique Chantrel, Amélie Cournoyer, Marco Fugazza, Poul Hansen, Jan Hoffmann, Tomasz Kulaga, Anila Premti, Luisa Rodríguez, Benny Salo, Kamal Tahiri, Hidenobu Tokuda, Pamela Ugaz and Frida Youssef were contributing authors.

The report benefitted from reviews and contributions by officials from the International Maritime Organization, the International Labour Organization partners of the TrainForTrade Port Management Programme and the five regional commissions of the United Nations (ECA, ECE, ECLAC, ESCAP, and ESCWA): Julian Abril Garcia, Peter Adams, Mario Apostolov, Yarob Badr, Jan de Boer, Aicha Cherif, Ismael Cobos Delgado, Yann Duval, Martina Fontanet Solé, Fouad Ghorra, Fredrik Haag, Robert Lisinge, Dorota Lost-Sieminska, Ricardo Sanchez, Lynn Tan, Lukasz Wyrowski and Brandt Wagner.

Comments and suggestions from the following reviewers are gratefully acknowledged: Hashim Abbas Syed, Roar Adland, Stefanos Alexopoulos, Jason Angelopoulos, Tracy Chatman, Trevor Crowe, Neil Davidson, Juan Manuel Díez Orejas, Mahin Faghfouri, Mike Garrat, Nadia Hasham, Joe Hiney, Julian Hoffmann Anton, Onno Hoffmeister, Roel Janssens, Lars Jensen, Björn Klippel, Eleni Kontou, Juan Manuel, Antonis Michail, Turloch Mooney, Richard Morton, Plamen Natzkoff, Jean-Paul Rodrigue, Peter Sand, Torbjorn Rydbergh, Alastair Stevenson, Stelios Stratidakis, Christa Sys, Antonella Teodoro and Ruosi Zhang. Experts from the International Chamber of Shipping reviewed chapter 2.

Comments received from UNCTAD divisions as part of the internal peer review process, as well as comments from the Office of the Secretary-General, are acknowledged with appreciation.

The *Review* was edited by Peter Stalker. Administrative, editing, and proofreading support was provided by Wendy Juan. Magali Studer designed the publication, and Juan Carlos Korol did the formatting.

Special thanks are also due to Vladislav Shuvalov for reviewing the publication in full.

TABLE OF CONTENTS

Acknowledgements iii
Abbreviations ix
Note xii
Overview xiv

1. **International maritime trade and port traffic** 1
   - A. Volumes of international maritime trade and port traffic 3
   - B. Outlook and longer-term trends 19
   - C. Policy considerations and action areas 23

2. **Maritime transport and infrastructure** 29
   - A. The world fleet 31
   - B. Shipping companies and operations: adapting maritime transport supply in an uncertain environment 42
   - C. Port services and infrastructure supply 46
   - D. The Impact of COVID-19 on ports: lessons from the UNCTAD TrainForTrade Port Management Programme 49
   - E. Summary and policy considerations 54

3. **Freight rates, maritime transport costs and their impact on prices** 57
   - A. Record-breaking container freight rates 59
   - B. Dry bulk freight rates also reach highs 64
   - C. Tanker freight rates dip to the lowest levels ever 65
   - D. Economic impact of high container freight rates, particularly in smaller countries 66
   - E. Structural determinants of maritime transport costs 70
   - F. Summary and policy considerations 74
   - Technical Notes 78

4. **Key performance indicators for ports and the shipping fleet** 87
   - A. Port calls and turnaround times 89
   - B. Liner shipping connectivity 93
   - C. Port cargo handling performance 99
   - E. Greenhouse gas emissions by the world fleet 105
   - F. Summary and policy considerations 106

5. **The COVID-19 seafarer crisis** 109
   - A. Seafarers crisis – recent developments 111
   - B. Seafarer crisis – implementation of the ILO Maritime Labour Convention, 2006, as amended (MLC 2006) 115
   - C. Crew changes and key worker status – other relevant international legal instruments 117
   - D. The way forward 119

6. **Legal and regulatory developments and the facilitation of maritime trade** 125
   - A. Technological developments in the maritime industry 127
   - B. Regulatory developments relating to international shipping, climate change and other environmental issues 128
   - C. Legal and regulatory implications of the COVID-19 pandemic 133
   - D. Other legal and regulatory developments affecting transportation 133
   - E. Maritime transport within the WTO Trade Facilitation Agreement 135
   - F. FAL Convention 139
   - G. ASYCUDA ASYHUB case studies 141
   - H. Summary and policy considerations 142

Tables

1	World fleet by principal vessel type, 2020–2021	xvi
2	Five largest seafarer-supplying countries 2021 supplying countries 2021	xx
1.1	International maritime trade, 1970–2020	3
1.2	International maritime trade 2019–2020, by type of cargo, country group and region	4
1.3	World economic growth, 2019–2021	6
1.4	Growth in the volume of world merchandise trade, 2019–2021	7
1.5	Tanker trade, 2019–2020	11
1.6	Dry bulk trade 2019–2020	12
1.7	Major dry bulk and steel: producers, users, exporters, and importers, 2020	13
1.8	Containerized trade on East-West trade routes, 2016–2020	15
1.9	Containerized trade on major East-West trade routes, 2014–2021	15
1.10	World container port throughput by region, 2019–2020	17
1.11	International maritime trade developments forecasts, 2021–2026	19
2.1	World fleet by principal vessel type, 2020–2021	31
2.2	Age distribution of world merchant fleet by vessel type, 2021 and average age 2020–2021	32
2.3	Top 25 ship-owning economies, as of 1 January 2021	35
2.4	Ownership of the world fleet, ranked by carrying capacity in dead-weight tons, 2021	36
2.5	Leading flags of registration by dead-weight tonnage, 2021	38
2.6	Leading flags of registration, ranked by value of total tonnage, 2021 (million US dollars) and principal vessel types	39
2.7	Deliveries of newbuildings by major vessel types and countries of construction, 2020	39
2.8	Reported tonnage sold for ship recycling by major vessel type and country of ship recycling, 2020	41
2.9	Status of uptake of selected technologies in global shipping, as of 14 June 2021	42
2.10	Some proposed IMO measures to reduce greenhouse gas emissions	43
2.11	World fleet by fuel type as of 1 January 2021	45
2.12	Industrial port projects capitalizing on green opportunities to generate new revenue streams	48
2.13	Factors affecting the development of smart green ports	49
2.14	Port Performance Scorecard indicators, 2016–2020	50
3.1	Contract freight rates, inter-regional, 2018–2020, $ per 40-foot container	62
4.1	Time in port, age, and vessel sizes, by vessel type, 2020, world total	90
4.2	Port calls and median time spent in port, container ships, 2020, top 25 countries	91
4.3	Top 25 ports under the World Bank IHS Markit Container Port Performance Index 2020	99
4.4	Minutes per container move, by range of call size, top 25 countries by port calls	101
4.5	Cargo and vessel handling performance for dry bulk carriers. Top 30 economies by vessel arrivals, average values for 2018 to first half of 2021	103
4.6	Cargo and vessel handling performance for tankers. Top 30 countries by vessel arrivals, average values for 2018 to first half of 2021	104
5.1	Neptune Declaration Crew Change Indicator, July 2021	113
5.2	Five largest seafarer-supply countries, 2021	115
6.1	Key performance indicators of the Kenya Trade Information Portal	138

Figures

1	International maritime trade, world gross domestic product (GDP) and maritime trade-to-GDP ratio, 2006 to 2021	xii
2	Simulated impact of current container freight rate surge on import and consumer price levels	xv
3	Median time in port, number of port calls, and maximum vessel sizes, by country, container ships, 2020	xvii
1.1	International maritime trade, world gross domestic product (GDP) and maritime trade-to-GDP ratio, 2006 to 2021	5
1.2	Participation of developing countries in international maritime trade, selected years	5
1.3	International maritime trade, by region, 2020	5
1.4	International maritime trade by cargo type, selected years	8
1.5	International maritime trade in cargo ton-miles, 2001–2021	9
1.6	World capesize dry bulk trade by exporting region in tons and ton-miles, 2019–2020	10
1.7	World ultra-large tanker trade by exporting region in ton and ton-miles, 2018–2020	10
1.8	Global containerized trade, 1996–2021	14
1.9	Global containerized trade by route, 2020	14
1.10	World container port throughput by region, 2019–2020	18
1.11	Leading 20 global container ports, 2019–2020	18
2.1	Annual growth rate of world fleet, dead-weight tonnage, 2000–2020	31
2.2	Age distribution of the global fleet, share of the global carrying capacity, 2012–2021	33
2.3	Age distribution of the fleet, as at beginning of 2021, per development status groups	33
2.4	Share of mega-vessels in the global container ship fleet carrying capacity by TEU, 2011–2021	34
2.5	Number of mega-containerships	34
2.6	Mega-vessel distinct journeys through the Panama and Suez canals, daily averages, from 2012 until 4 June 2021	34
2.7	Live and on-order global fleet by ship type	37
2.8	Growth of world fleet orderbook, 2012–2021, percentage change in dead-weight tonnage	40
2.9	World tonnage on order, selected ship types, 2000–2021	41
2.10	Percentage change in cost intensity by ship segment, average size and median distance travelled	44
2.11	Cargo and revenue, 2016–2020	51
2.12	Average revenue mix of ports, 2016–2020	52
3.1	Growth of demand and supply in container shipping, 2007–2021, percentage	59
3.2	CCFI composite index, 2011-2021 (quarterly)	60
3.3	Shanghai Containerized Freight Index weekly spot rates, 1 July 2011 to 30 July 2021, selected routes	60
3.4	New ConTex index, July 2011–July 2021	63
3.5	Baltic Exchange Dry Index, January 2010–July 2021	65
3.6	Average weighted earnings all bulkers ($/day), July 2001–July 2021	65
3.7	Average earnings, all tankers, July 2011–July 2021	66
3.8	Simulated impact of current container freight rate surge on import and consumer price levels	67

3.9	Simulated impacts of the container freight rate surge on consumer price levels, by country and by product	68
3.10	Simulated impacts of container freight rate surges on prices for importers, consumers and firms, global average	69
3.11	Simulated impact of container freight rate surges on production costs, by country and size of economy	69
3.12	Simulated dynamic impacts of container freight rate increase on industrial production	70
3.13	Transport costs for importing goods by transport mode, world, LDCs, and LLDCs, 2016, percentage of FOB value	71
3.14	Transport costs heatmap for importing goods, all modes of transport, 2016, percentage of FOB value	71
3.15	Maritime transport costs for importing goods and distances from trading partners	72
3.16	Maritime transport costs for importing goods, by country and size of economy	73
3.17	Impact of structural determinants on maritime transport costs for importing goods	73
3.18	Maritime transport costs by direction of the trade imbalance	74
3.19	Impacts of trade imbalance and trade volume on maritime transport costs	74
4.1	Port calls per half year, world total, 2018–2020	89
4.2	Port calls per half year, regional totals, 2018–2020	89
4.3	Container ship port calls and time in port, 2020	90
4.4	Container ship port calls and maximum ship sizes, 2020	91
4.5	Container ship port calls in Africa and time in port, 2020	92
4.6	Container ship port calls in Africa and maximum ship sizes, 2020	92
4.7	Median time in port, number of port calls, and maximum vessel sizes, per country, container ships, 2020	92
4.8	Liner shipping connectivity index, top 10 countries, first quarter 2006 to second quarter 2021	93
4.9	Port Liner Shipping Connectivity Index, top 10 ports as of second quarter 2021, first quarter 2006 to second quarter 2021	94
4.10	Liner Shipping Connectivity Index, country and port level, 2020	95
4.11	Trends in global container ship deployment, first quarter 2006 to second quarter 2021	96
4.12	Trends in vessel sizes and number of companies providing services, selected countries, first quarter 2006 to second quarter 2021	97
4.13	Relationship between maximum vessel sizes, deployed capacity, and the number of companies, second quarter 2021	98
4.14	Liner Shipping Bilateral Connectivity Index (LSBCI) and its components, first quarter 2006 to second quarter 2021	99
4.15	Minutes per container move for container ships, by range of port call size	100
4.16	Time in port (hours) for container ships, by range of port call size	100
4.17	Correlation between time in port (hours) and minutes per container move, all call sizes	101
4.18	Correlation between time in port (hours) and minutes per container move, only calls with 1001 to 1500 containers per call	101
4.19	Carbon dioxide emissions by vessel type, monthly, million tons, 2011–2021	105
4.20	Carbon dioxide emissions by flag state, annual, 2011–2020, million tons	106

Boxes

1	Implications of AfCFTA for maritime transport in Africa	20
2.1	Divided views on whether oil should be replaced by LNG	46
2.2	Building port resilience UNCTAD experience	46
2.3	Guidance and standards for intermodal operations	47
2.4	Port performance analysis of the Port of Gijon in 2020	51
2.5	Port performance analysis of the national port system in Peru in 2020	52
2.6	Gender and development in the Philippine Ports Authority and its journey	53
3.1	Impact of COVID-19 on maritime freight rates in the Arab region	61
4.1	Port performance in Latin America and the Caribbean – differences between types of terminals	102
5.1	The case of the Philippines	114
6.1	The Framework Agreement on Facilitation of Cross-Border Paperless Trade in Asia and the Pacific - Maritime implications	138
6.2	IMO Compendium on Facilitation and Electronic Business	139
6.3	Components of the Digitizing Global Maritime Trade project	141
6.4	Customs formalities concerning entry or exit	142

ABBREVIATIONS

AfCFTA	African Continental Free Trade Area
AGTC	European Agreement on Important International Combined Transport Lines
APEC	Asia-Pacific Economic Cooperation
ASYCUDA	Automated System for Customs Data
ASYHUB	ASYCUDA data integration system
B2B	business to business
B2G	business to government
BIMCO	Baltic and International Maritime Council
CAPEX	capital expenditure
CCFI	China Containerized Freight Index
CIF	cost, insurance and freight
CII	Carbon Intensity Indicator
CO_2	carbon dioxide
CPPI	Container Port Performance Index
DGMT	Digitizing Global Maritime Trade
dwt	deadweight tonnage
EBITDA	earnings before interest, taxes, depreciation and amortization
ECA	Economic Commission for Africa
ECE	United Nations Economic Commission for Europe
ECLAC	United Nations Economic Commission for Latin America and the Caribbean
EEDI	Energy Efficiency Design Index
EEXI	Energy Efficiency Existing Ship Index
ESCAP	United Nations Economic Commission for Asia and the Pacific
ESCWA	United Nations Economic and Social Commission for Western Asia
eSW	electronic single window
eTIR	electronic International Road Transport system
EU	European Union
FAL Convention	Convention Facilitation of International Maritime Traffic
FIATA	International Federation of Freight Forwarders Associations
FOB	free on board

G2B	government to business
GAD	gender and development
GDP	Gross domestic product
GT	Gigaton
GTCDIT	Global Transport Costs Dataset for International Trade
GVC	global value chain
HFO	heavy fuel oil
ICAO	International Civil Aviation Organization
ICS	Institute Of Chartered Shipbrokers
IFO	intermediate fuel oil
ILO	International Labour Organization
IMF	International Monetary Fund
IMO	International Maritime Organisation
IOM	International Organization for Migration
IOPC FUNDS	International Oil Pollution Compensation Funds
IRU	International Road Transport Union
ISM	International Safety Management
ISO	International Standards Organization
ISPS	International Ship and Port Facility Security
ITF	International Transport Workers' Federation
ITS	intelligent transport systems
kw	kilowatt
LDC	least developed country
LLDC	landlocked developing country
LNG	liquified natural gas
LPG	liquified petroleum gas
MARPOL Convention	International Convention for the Prevention of Pollution from Ships
MASS	maritime autonomous surface ship
MDH	Maritime Declaration of Health
MDO	marine diesel oil
MEPC	IMO Marine Environment Protection Committee
MGO	marine gasoil

MLC	Maritime Labour Convention
MMT-RDM	Multi-Modal Transport Reference Data Model
MNSW	maritime national single window
MSC	IMO Maritime Safety Committee
MSW	maritime single window
NTFC	National Trade Facilitation Committee
OECD	Organisation for Economic Co-operation and Development
OPEC	Organization of the Petroleum Exporting Countries
PCS	port community system
PHEIC	public health emergency of international concern
PIANC	World Association for Waterborne Transport Infrastructure
PPA	Philippine Ports Authority
PPPs	public-private partnerships
PPS	Port Performance Scorecard
R&D	research and development
SCFI	Shanghai Containerized Freight Index
SID	Seafarers' Identity Document
SIDS	small island developing states
STCW	Standards of Training, Certification and Watchkeeping for Seafarers
TEU	twenty-foot-equivalent unit
TIP	Trade Information Portal
UN/CEFACT	The United Nations Centre for Trade Facilitation and Electronic Business
UNCITRAL	United Nations Commission on International Trade Law
UNCTAD	United Nations Conference on Trade and Development
UNDESA	UN Department of Economic and Social Affairs
UNFCCC	United Nations Framework Convention on Climate Change
UNOHRLLS	United Nations Office of the High Representative for the Least Developed Countries, Landlocked Developing Countries and Small Island Developing States
VLSFO	very low sulphur fuel oil
WCO	World Customs Organization
WHO	World Health Organization
WIOD	World Input-Output Database
WTO	World Trade Organization

NOTE

The *Review of Maritime Transport* is a recurrent publication prepared by the UNCTAD secretariat since 1968 with the aim of fostering the transparency of maritime markets and analysing relevant developments. Any factual or editorial corrections that may prove necessary, based on comments made by Governments, will be reflected in a corrigendum to be issued subsequently.

This edition of the *Review* covers data and events from January 2020 until June 2021. Where possible, every effort has been made to reflect more recent developments.

All references to dollars ($) are to United States dollars, unless otherwise stated.

"Ton" means metric ton (1,000 kg) and "mile" means nautical mile, unless otherwise stated.

Because of rounding, details and percentages presented in tables do not necessarily add up to the totals.

Two dots (..) in a statistical table indicate that data are not available or are not reported separately.

All websites were accessed in September 2021.

The terms "countries" and "economies" refer to countries, territories or areas.

Since 2014, the *Review of Maritime Transport* does not include printed statistical annexes. UNCTAD maritime statistics are accessible via the following links:

    All datasets: http://stats.unctad.org/maritime

    Merchant fleet by flag of registration: http://stats.unctad.org/fleet

    Share of the world merchant fleet value by flag of registration: http://stats.unctad.org/vesselvalue_registration

    Merchant fleet by country of ownership: http://stats.unctad.org/fleetownership

    Share of the world merchant fleet value by country of beneficial ownership: http://stats.unctad.org/vesselvalue_ownership

    Ship recycling by country: http://stats.unctad.org/shiprecycling

    Shipbuilding by country in which built: http://stats.unctad.org/shipbuilding

    Seafarer supply: http://stats.unctad.org/seafarersupply

    Liner shipping connectivity index: http://stats.unctad.org/lsci

    Liner shipping bilateral connectivity index: http://stats.unctad.org/lsbci

    Container port throughput: http://stats.unctad.org/teu

    Port liner shipping connectivity index: http://stats.unctad.org/plsci

    Port call performance (Time spent in ports, vessel age and size), annual: http://stats.unctad.org/portcalls_detail_a

    Port call performance (Time spent in ports, vessel age & size), semi-annual: http://stats.unctad.org/portcalls_detail_sa

    Number of port calls, annual: http://stats.unctad.org/portcalls_number_a

    Number of port calls, semi-annual: http://stats.unctad.org/portcalls_number_sa

    Seaborne trade: http://stats.unctad.org/seabornetrade

    National maritime country profiles: http://unctadstat.unctad.org/CountryProfile/en-GB/index.html

Vessel groupings used in the *Review of Maritime Transport*

Group	Constituent ship types
Oil tankers	Oil tankers
Bulk carriers	Bulk carriers, combination carriers
General cargo ships	Multi-purpose and project vessels, roll-on roll-off cargo ships, general cargo ships
Container ships	Fully cellular container ships
Other ships	Liquefied petroleum gas carriers, liquefied natural gas carriers, parcel (chemical) tankers, specialized tankers, refrigerated container ships, offshore supply vessels, tugboats, dredgers, cruise, ferries, other non-cargo ships
Total all ships	Includes all the above-mentioned vessel types

Approximate vessel-size groups according to commonly used shipping terminology

Crude oil tankers

Ultralarge crude carrier	320,000 dead-weight tons (dwt) and above
Very large crude carrier	200,000–319,999 dwt
Suezmax crude tanker	125,000–199,999 dwt
Aframax/longe-range 2 crude tanker	85,000–124,999 dwt
Panamax/long-range 1 crude tanker	55,000–84,999 dwt
Medium-range tankers	40,000–54,999 dwt
Short-range/Handy tankers	25,000–39,000 dwt

Dry bulk and ore carriers

Capesize bulk carrier	100,000 dwt and above
Panamax bulk carrier	65,000–99,999 dwt
Handymax bulk carrier	40,000–64,999 dwt
Handysize bulk carrier	10,000–39,999 dwt

Container ships

Neo-Panamax	Container ships that can transit the expanded locks of the Panama Canal with up to a maximum 49 m beam and 366 m length overall; fleets with a capacity of 12,000–14,999 20-foot equivalent units (TEUs) include some ships that are too large to transit the expanded locks of the Panama Canal based on current dimension restrictions.
Panamax	Container ships above 3,000 TEUs with a beam below 33.2 m, i.e., the largest size vessels that can transit the old locks of the Panama Canal.
Post Panamax	Fleets with a capacity greater than 15,000 TEUs include some ships that are able to transit the expanded locks.

Source: Clarksons Research.

Note: Unless otherwise indicated, the ships mentioned in the *Review of Maritime Transport* include all propelled seagoing merchant vessels of 100 gross tons and above, excluding inland waterway vessels, fishing vessels, military vessels, yachts, and fixed and mobile offshore platforms and barges (with the exception of floating production storage, offloading units and drillships).

OVERVIEW

Maritime transport defied the COVID-19 disruption. In 2020, volumes fell less dramatically than expected and by the end of the year had rebounded, laying the foundations for a transformation in global supply chains and new maritime trade patterns

The COVID-19 pandemic disrupted maritime transport, though the outcome was less damaging than initially feared. The shock in the first half of 2020 caused maritime trade to contract by 3.8 per cent in the year 2020. But in the second half of the year there was a nascent, if asymmetric, recovery, and by the third quarter, volumes had returned, for both containerized trade and dry bulk commodities. However, there has yet to be a full recovery for tanker shipping.

Maritime trade has performed better than expected partly because the COVID-19 pandemic unfolded in phases and at different speeds, with diverging paths across regions and markets. The rebound in trade flows was also the result of large stimulus packages, and increased consumer spending on goods, with a growth in e-commerce, especially in the United States. Later, there was more general optimism in advanced regions from the rollout of vaccines. But it was also partly due to unlocking pent-up demand for cars, for example, and to restocking and inventory-building. The rebound was fairly swift because, unlike the global financial crisis of 2009, the downturn was not synchronized across the world.

In 2021, in tandem with the recovery in merchandise trade and world output, maritime trade is projected to increase by 4.3 per cent (figure 1). The medium-term outlook also remains positive, though subject to mounting risks and uncertainties, and moderated in line with projected lower growth in the world economy. Over the past two decades, compound annual growth in maritime trade has been 2.9 per cent, but over the period 2022–2026, UNCTAD expects that rate to slow to 2.4 per cent.

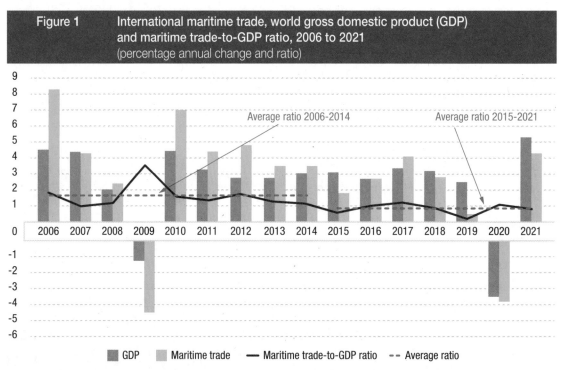

Figure 1 International maritime trade, world gross domestic product (GDP) and maritime trade-to-GDP ratio, 2006 to 2021
(percentage annual change and ratio)

Source: UNCTAD calculations, based on the *Review of Maritime Transport*, various issues, data from UNCTADstat and table 1.1 of the UNCTAD *Trade and Development Report 2021. From recovery to resilience: The development dimension.*

Maritime transport navigated through the pandemic, but there was an unprecedented humanitarian crisis for seafarers

While carriers generally managed to mitigate the shock and disruption, port and landside operations found it more difficult to adjust, and seafarers were in a precarious situation as the pandemic triggered an unprecedented global crew-change crisis. The health risks and related travel restrictions meant that hundreds of thousands of seafarers could not return home, while an equivalent number were unable to join their ships and to provide for their families.

Hardest hit has been tanker shipping, but the impact has been less for containerized trade, gas shipments, and dry bulk commodities

Lockdowns, travel restrictions and production cuts have compressed the demand for fuel. In 2020, shipments of crude oil, refined petroleum products, and gas together fell by 7.7 per cent. The impact was less, however, for dry bulk commodity trade: supported by strong demand from China for iron ore and grain, total dry bulk trade fell by only 1.5 per cent. Containerized trade also resisted, falling by only 1.1 per cent. Global container port throughput fell at a roughly similar rate – and in 2020 totalled 815.6 million twenty-foot equivalent units (TEU).

Logistical bottlenecks, and soaring costs, along with an asymmetric recovery, have heightened uncertainty

Maritime trade weathered the storm in 2020 and the short-term outlook remains positive. However, the emerging multi-paced recovery is inherently fragile as many countries and regions continue to lag. In addition to new pandemic risks and the dangers of a two-track vaccination pattern where developing countries continue to fall behind, other risks are casting a shadow on the recovery. While not all countries have been able to deploy large stimuli packages and support measures, an untimely ending of the existing support measures in advanced economies could potentially stifle growth and hinder the nascent recovery. The pandemic's impacts and legacies are likely to linger and the future shape and contours of the next normal for the world economy remain uncertain.

The nascent recovery has also been hindered by supply-chain bottlenecks. The rebound in trade, combined with pandemic-induced restrictions in logistics operations has led to shortages in equipment and containers, along with less reliable services, congested ports and longer delays and dwell times. For shipping, on the other hand, soaring freight rates, surcharges and fees have bolstered profitability.

Freight rates increased further following the March 2021 closure of the Suez Canal. The grounding of the 20,150-TEU container ship Ever Given blocked the canal, delaying ships heading for Europe, and increasing the constraints on ship and port capacity. Some voyages had to be re-routed around the Cape, adding up to 7,000 miles to the distance.

Whether the recovery lasts will depend critically on the path of the pandemic. Fresh waves of infection, combined with low vaccination rates, especially in developing countries, have led to new lockdowns and border closures. A broad-based recovery hinges to a large extent on a worldwide vaccine rollout. The International Monetary Fund estimates that $50 billion are required to end the pandemic and roll out vaccines across developing countries. This would bring not just health but also economic benefits since it would be tantamount to a large scale economic stimulus package that could accelerate economic recovery and by 2025 generate some $9 trillion in additional global output.

Seafarers are increasingly being recognized as "key workers" who are keeping shipping and trade moving, while also being at the front line of the health crisis. Since seafarers come predominantly from developing regions, industry and government should move quickly to implement vaccine procurement and distribution plans.

The longer-term outlook is being reshaped by structural megatrends that transcend the pandemic and its immediate impact

Eventually, the logistical hurdles caused by large swings in demand could dissipate as global trade patterns normalize. However, the pandemic has also accelerated megatrends that in the longer-term could transform the maritime transport landscape.

By exposing the vulnerabilities of existing supply chains, the COVID-19 disruption has sharpened the need to build resilience. COVID-19 emphasized the importance of ensuring continuity in supply chains and the need for them to become more resilient, responsive, and agile.

Discussions over the future of globalization have ushered calls to take a fresher look at the configuration of the extended supply chains to reduce heavy reliance on distant suppliers. Some are arguing that reshoring and nearshoring will accelerate, resulting in deep reconfiguration of supply chains. While the structural trends that had emerged over a decade ago and accelerated during recent trade tensions are likely to result in changes to globalization patterns and features, an outright end to globalization *per se* is unlikely.

It may be fairly straightforward to reshore labour-intensive and low-value production, but it is more complex to move production and switch suppliers for mid-and high-value-added manufacturing. Instead, enterprises are likely to blend local and global sourcing, modifying their strategies according to product and geography

– with a blend of reshoring, diversification, replication, and regionalization. Nevertheless, for the near future China is likely to remain a leading manufacturing site. Automation could make reshoring and nearshoring more economically viable in the longer term. Hybrid operating models involving just-in-time (i.e., material moved just before its use in the manufacturing process) and just-in-case (i.e., where companies keep large inventories to minimize stocks being sold out) supply chain models are likely to emerge. Combined, these trends will change distances and routes, increasing the need for more flexible shipping services. They also entail implications for vessel types and sizes, ports of call, and distance travelled.

The pandemic has accelerated pre-existing digitalisation and environmental sustainability trends. Technological advances have enabled shipping and ports to continue operations while minimizing interaction and physical contact. New technologies have also stimulated the rise of online commerce which has transformed consumer shopping habits and spending patterns. The growth in online trade has increased the demand for distribution facilities and warehousing that are digitally enabled and offer value-added services. All these developments are expected to generate new business opportunities for shipping and ports as well as for other players in the maritime supply chain.

Technology will also be critical for advancing environmental sustainability. While designing their stimulus packages and post-pandemic plans, many governments aim to harness the synergies between technology, environmental protection, efficiency, and resilience. Businesses and governments recognize that adapting to the post-pandemic world and building back better requires adding economic, social and environmental value and creating new business opportunities, not least for maritime transport.

Supply not keeping pace with demand

In 2020, the global commercial shipping fleet grew by 3 per cent, reaching 99,800 ships of 100 gross tons and above. By January 2021, capacity was equivalent to 2,13 billion dead weight tons (dwt) (table 1). During 2020, delivery of ships declined by 12 per cent, partly due to lockdown-induced labour shortages that disrupted marine-industrial activity. The ships delivered were mostly bulk carriers, followed by oil tankers and container ships. As owners and operators tried to cope with tight vessel supply, they were also buying more second-hand ships with a resulting increase in prices. Recycling rates also increased in 2020, although compared to previous years, the levels remain low.

During 2020, orders for new ships had declined by 16 per cent, continuing a downward trend observed in previous years. In early 2021, however, shipping companies reacted to the capacity constraints with a surge of new orders, especially for container ships for which orders were the highest for the last two decades. There were also more orders for LNG carriers.

Table 1 — World fleet by principal vessel type, 2020–2021
(thousand dead-weight tons and percentage)

Principal types	2020		2021		Percentage change 2021 over 2020
Bulk carriers	879 725	42.47%	913 032	42.77%	3.79%
Oil tankers	601 342	29.03%	619 148	29.00%	2.96%
Container ships	274 973	13.27%	281 784	13.20%	2.48%
Other types of ships:	238 705	11.52%	243 922	11.43%	2.19%
Offshore supply	84 049	4.06%	84 094	3.94%	0.05%
Gas carriers	73 685	3.56%	77 455	3.63%	5.12%
Chemical tankers	47 480	2.29%	48 858	2.29%	2.90%
Other/not available	25 500	1.23%	25 407	1.19%	-0.36%
Ferries and passenger ships	7 992	0.39%	8 109	0.38%	1.46%
General cargo ships	76 893	3.71%	76 754	3.60%	-0.18%
World total	**2 071 638**		**2 134 640**		**3.04%**

Source: UNCTAD calculations, based on data from Clarksons Research.
Note: Propelled seagoing merchant vessels of 100 tons and above; beginning-of-year figures.

During the second half of 2020, and into 2021, world trade gradually recovered but supply was less elastic and constrained by COVID-19 related delays and congestion – leading to a significant increase in container freight rates.

The future demand/supply balance will also be impacted by regulatory requirements to align shipping operations with decarbonization targets. Introduced under the auspices of the International Maritime Organization (IMO), these new regulations will require replacing some of the existing fleet so will entail significant costs. As well as creating a degree of uncertainty, this could reduce the capital available to expand the fleet to cater for trade growth.

Cost pressure and soaring rates and surcharges would weigh on smaller players and prices

Since the second half of 2020 there has been an increase in freight rates. While demand for containerized goods has been higher than expected, shipping capacity has been constrained by logistical hurdles and bottlenecks and shortages in container shipping equipment. Unreliable schedules, and port congestion have also led to a surge in surcharges and fees, including demurrage and detention fees.

These soaring costs are a challenge for all traders and supply chain managers, but especially for smaller shippers who, compared with the larger players, may be less able to absorb the additional expense and are at a disadvantage when negotiating rates and booking space on ships. Smaller shippers and low-value paying cargo may thus find it difficult to secure service contracts and could see their margins eroded.

Freight rates are expected to remain high. Demand is strong and there is growing uncertainty on the supply side, with concerns about the efficiency of transport systems and port operations. In the face of these cost pressures and lasting market disruption, it is increasingly important to monitor market behaviour and ensure transparency when it comes to setting rates, fees, and surcharges. There have been calls for governments to intervene, and for regulators to apply closer oversight and address unfair market practices.

If sustained, the current surge in container freight rates, will significantly increase both import and consumer prices. UNCTAD's simulation model suggests that global import price levels will increase on average by 11 per cent as a result of the freight rate increases (figure 2). Hardest hit will be the small island developing states (SIDS) who depend for their merchandise imports primarily on maritime transport and who are simulated to face a cumulative increase of 24 per cent with a time lag of about a year.

Higher container freight rates will also have a sizeable impact on consumer prices. If container freight rates remain at their current high levels, then in 2023 global consumer prices are projected to be 1.5 per cent higher than they would have been without the freight rate surge. The impact is expected to be more significant for smaller economies that depend heavily on imported goods for much of their consumption needs. In SIDS, the cumulative increase in consumer prices is expected to be 7.5 per cent and in the Least Developed Countries (LDCs) 2.2 per cent.

Figure 2 Simulated impact of current container freight rate surge on import and consumer price levels

Sources: Based on data provided by Clarksons Research, *Shipping Intelligence Network*, the International Monetary Fund, *International Financial Statistics* and *Direction of Trade Statistics*, UNCTADstat, and the World Bank, *World Integrated Trade Solution* and *Commodity Price Data (The Pink Sheet)*.

Note: The impact of container freight rate surges on prices is assessed based on a 243 per cent increase in the China Containerized Freight Composite Index between August 2020 and August 2021. The simulation model assumes that freight rates in August 2021 will be sustained over the remaining simulation period (September 2021 to December 2023) and all other factors are held constant over the entire simulation period (August 2020 to December 2023).

Some goods will be affected more than others by the surge in container freight rates. Most exposed are goods manufactured through integrated supply chains. Globalized production processes entail a greater use of shipping, with intermediate goods often crossing borders multiple times within and between regions. This is the case, for example, for East Asian goods destined for major markets in North America and Europe. For computers, and electronic and optical products, for example, the consumer price uplift induced by the current freight rate surge could be 11 per cent.

Higher shipping costs will also affect some low-value-added products: for furniture, for example, and textiles, garments and leather products, the consumer price uplifts could be ten per cent. These increases could erode the competitive advantages of smaller economies that produce many of these goods. At the same time, these countries will find it more difficult to import the high-technology machinery and industrial materials they need to move up the value chain, diversify their economies and achieve the Sustainable Development Goals (SDGs).

Even in major economies, lingering high container freight rates and disruption in maritime transport in the short- to medium-term threaten to undermine recovery. UNCTAD's analysis concludes that in the United States and the euro area, for example, a 10 per cent increase in container freight rates could lead to a cumulative contraction in industrial production of around 1 per cent.

Structural factors keep maritime transport costs higher in developing regions

The current historical highs in freight rates are largely driven by pandemic-induced shocks and unexpected upward swings in shipping demand. But in the longer term, shipping and port prices are driven by structural factors such as port infrastructure, economies of scale, trade imbalances, trade facilitation, and shipping connectivity – all of which have lasting impacts on maritime transport costs and trade competitiveness. An analysis based on a new UNCTAD-World Bank transport costs dataset, shows that significant structural improvements could reduce maritime transport costs by around four per cent. Interventions and policies that address the structural determinants of maritime transport costs can thus help mitigate the impacts from cyclical factors and disruptions.

Other structural issues that will increase prices include the new regulations on decarbonizing shipping. The recently adopted IMO short-term measure on greenhouse gas reduction is expected to reduce average shipping speeds and increase maritime transport costs, especially for developing countries, and in particular the SIDS.

COVID-19 slows operations for ships and ports

In the first half of 2020, reflecting the slump in shipping demand, cargo-carrying ships made fewer port calls. The number of calls subsequently increased, particularly in Europe, East Asia, and South-Eastern Asia, albeit not yet to pre-pandemic levels.

In 2020, terminal operators, authorities, and intermodal transport providers took measures to contain COVID-19 and, as a result, ships had to spend more time in ports that were operating more slowly. The greatest delays were for dry break bulk carriers for which cargo operations tend to be less automated and more labour-intensive so were slowed by measures to reduce social contact.

Turnaround times can differ significantly between countries (figure 3). One group of countries with faster turnarounds comprises those with fewer arrivals and only small ships and with only few containers loaded and unloaded during each port call. These include Dominica, Saint Kitts and Nevis, and Saint Vincent and the Grenadines. Another group with fast turnarounds comprises those that have the latest port technologies and infrastructure and can accommodate the largest container vessels; they benefit from economies of scale and thus tend to attract the highest number of port calls. These include Japan, Hong Kong China, and Taiwan Province of China. Efficient ports initiate a positive feed-back loop: high efficiency makes their ports attractive as ports of call, further boosting the number of arrivals. Countries in the middle of the distribution report a wide range of median port waiting times, reflecting differences in efficiency and other variables such as vessel age and cargo throughput.

Shipping and port performance is generally lower in developing countries. They have higher transport costs and lower connectivity because they are often further away from their overseas markets and are hampered by diseconomies of scale and lower levels of digitalization.

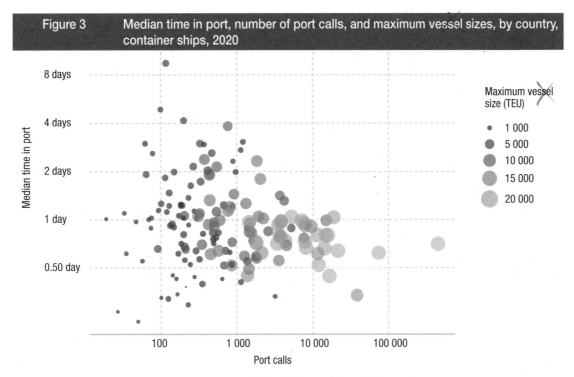

Figure 3 Median time in port, number of port calls, and maximum vessel sizes, by country, container ships, 2020

Source: UNCTAD, based on data provided by MarineTraffic. Both axes in logarithmic scale.
Note: Ships of 1,000 GT and above. For the complete table of countries, see http://stats.unctad.org/maritime.

Positive trends in port governance and gender participation

Each year, UNCTAD uses data from its TrainForTrade Port Management Programme to benchmark countries against each other using the Port Performance Scorecard (PPS). Many other port performance projects focus on service provision such as cargo handling, but the PPS, which uses data for 26 indicators, enables comparisons between entire port entities, providing data that is valuable for strategic planning within ports and for evidence-based policy analysis at regional and state levels.

Member ports' annual throughput in 2020 ranged from 1.5 million to 80.9 million tonnes. Around half of were in the smallest category, less than 5 million tonnes, and the medium category, 5 million to 10 million tonnes, a range of volumes that was similar across all regions.

Since 2015, one of the six main categories in the PPS scorecard has been the rate of female participation in the port workforce. In 2019 and 2020, this remained low, at around 18 per cent. The rate was significantly higher in Europe at 25 per cent, though even here roles are not equally distributed between men and women. Women tend to be better represented in management and administrative roles, for which between 2019 and 2020 the proportion of women increased from 38 to 42 per cent. In this case, Asian members were above average at 52 per cent compared with those in Europe at 39 per cent. Women are far less likely to be working in cargo handling port operations. These results highlight the need for strategic policy interventions to deliver on Sustainable Development Goal 5 to "Achieve gender equality and empower all women and girls."

Port and shipping performance depend on trade and transport facilitation

Efficient maritime transport depends on effective trade and transport facilitation that reduces the time and cost of customs and other trade procedures and integrate new technologies for administrative formalities. Boosts the performance of the entire supply chain with positive effects on maritime transport.

The need for cross-border trade facilitation was highlighted by the COVID-19 pandemic, particularly for trade in medical equipment, drugs and emergency goods such as vaccines and personal protection equipment (PPE) – which could be held up at ports by red tape or by slow clearance procedures to comply with regulatory requirements.

In recent years, the introduction of new technology in administrative processes has boosted efficiency along the logistics supply chain. This has involved digitalization and automation of customs processes, paperless formalities, and the introduction of single-window services – the impetus for which was boosted during the COVID-19 pandemic.

An example of the use of ICT, is UNCTAD's Automated System for Customs Data (ASYCUDA) which involves automation and digitalization in supply chains. A recent development, the ASYHUB solution, smooths data transfer between ports of departure and arrival – using risk management concepts to help speed up clearance procedures and avoid goods being stuck in ports unnecessarily.

Another ICT innovation based on UNCTAD technology is the Trade Information Portal (TIP) – a website in each country that provides traders with easy access to information about trade regulations and procedures. The UNCTAD TIP offers importers and exporters online, step-by-step guides to trade-related procedures and also helps the country fulfil its obligations arising from the World Trade Organization Trade Facilitation Agreement. Today, 29 TIPs, based on UNCTAD technology, are being implemented globally by UNCTAD and the International Trade Centre. Results have been very positive. TIPs are most advanced in East Africa, where in Kenya, for example, greater transparency and simplification of a total of 52 trade procedures so far have reduced the time spent waiting in the queue, at the counter and in between steps by 110 hours, and the administrative fees for these 52 procedures by $482, i.e., about $11 per trade procedure on average.

Digitalization allows a paperless environment whereby trade procedures are all carried out online. For the traders this reduces time and cost and increases transparency and market access, while also reducing physical contact and the risks of contagion. In addition, smart digital solutions improve public administration of trade and boost efficiency in export, import and transit operations. Moreover, by minimizing the use of paper, trade facilitation can also help mitigate climate change.

Reforms in trade facilitation have been promoted by the multilateral trading system, particularly through the WTO Facilitation Agreement and the IMO Convention on Facilitation of International Maritime Traffic. These agreements provide common standards and regulations that have proved especially valuable during the COVID-19 pandemic. By providing governments with guidance and incentives for reforming trade facilitation, they have paved the way for further digitalization and enhanced transparency, and for rationalizing administrative formalities. These developments also promote robust public-private partnerships (PPPs), such as the National Trade Facilitation Committees and Port Community Systems that involve the business community in port operations. Efficient maritime trade and transport will depend on aligning and streamlining the mandates and work of the various PPPs.

A continuing crisis for seafarers stranded at sea

Globally there around 1.9 million seafarers working to facilitate the way we live. The BIMCO/ICS Seafarer Workforce Report 2021 estimated the global supply of seafarers at 1,892,720, up from 1,647,494 in 2015. Of these, 857,540 were officers, and 1,035,180 were ratings – the skilled seafarers who carry out support work. The five largest seafarer-supplying countries were the Philippines, the Russian Federation, Indonesia, China, and India, representing 44 per cent of the global workforce (table 2).

Table 2 Five largest seafarer-supplying countries 2021 supplying countries 2021

	All Seafarers	Officers	Ratings
1	Philippines	Philippines	Philippines
2	Russian Federation	Russian Federation	Russian Federation
3	Indonesia	China	Indonesia
4	China	India	China
5	India	Indonesia	India

Source: ISF and BIMCO, Seafarer Workforce Report 2021, London, 2021.

For the supplying countries seafarers are important sources of income. In 2019, the Philippines, for example, earned $30.1 billion from its overseas workers – 9.3 per cent of GDP and 7.3 per cent of gross national income (GNI) – of which $6.5 billion came from its seafarers. In 2020 total remittances fell 0.8 per cent to $29.9 billion, with those from seafarers falling 2.8 per cent to $6.4 billion.

During the COVID-19 pandemic, seafarers continued to demonstrate great professionalism and dedication, supporting the delivery of food, medical supplies, fuel, and other essential goods, and helping keep supply chains active and global commerce running.

However, hundreds of thousands of seafarers remain stranded at sea. Each month, crews need to be changed over – to prevent fatigue and comply with international maritime regulations for safety, health and welfare. Responding to COVID-19, governments closed many borders and imposed lockdowns and prohibited people from disembarking thus temporarily suspending crew changes. As a consequence, large numbers of seafarers have been unable to be replaced or repatriated after long tours of duty and had to extend their service on board. Even over a year into the pandemic, due to these restrictions, and the shortage of international flights, according to latest estimates by the International Chamber of Shipping, around 250,000 seafarers remain stranded, far beyond the expiration of their contracts. Yet, there is still no global consensus on uniform measures to allow for efficient crew changes and transfer.

During the pandemic, stakeholders, including international bodies, governments, and industry, have issued recommendations and guidance – aiming to ensure that seafarers are healthy and protected from COVID-19, have access to medical care, and are recognized as key workers and are vaccinated as a matter of priority, and also that ships and port facilities meet international sanitary requirements. Nevertheless, as the pandemic continues for a second year, seafarers remain very vulnerable.

With some notable exceptions, only a small proportion of the world's seafarers have been vaccinated. Belgium has demonstrated best practice, and July 2021 started a vaccination campaign for all seafarers arriving in a Belgian port, regardless of nationality.

To address seafarers' issues there has been a continuous level of cooperation among international organizations and industry bodies, including IMO, ILO, WHO, UNCTAD, ICS, and ITF, which have repeatedly expressed concern about the humanitarian crisis in the maritime shipping sector and urged Member States to designate seafarers and other marine personnel as key workers, accept seafarers' identity documents as evidence of their key worker status, and allow flexibility for ship owners and managers to divert ships to ports where crew change is possible without imposing penalties.

On 1 December 2020, the UN General Assembly unanimously adopted a resolution: International cooperation to address challenges faced by seafarers as a result of the COVID-19 pandemic to support global supply chains (A/RES/75/17). This urges Member States to designate seafarers and other marine personnel as key workers and encourages governments and other stakeholders to implement the "Industry Recommended Framework of Protocols for ensuring safe ship crew changes and travel during the Coronavirus (COVID-19) pandemic". It also calls upon governments to facilitate maritime crew changes – for example, by enabling them to embark and disembark, expediting travel and repatriation efforts, and ensuring access to medical care. The resolution also requests IMO, ILO and UNCTAD to inform the General Assembly at its 76th session on issues related to the resolution.

This follows earlier resolutions from other bodies. On 21 September 2020 the IMO's Maritime Safety Committee recommended action to facilitate ship crew change, access to medical care, and seafarer travel during the COVID-19 pandemic. According to IMO, as of the end of June 2021, 60 Member States and two Associate Members had signed on to designate seafarers as key workers. Similarly, on 8 December 2020 the Governing Body of the ILO, adopted the "Resolution concerning maritime labour issues and the COVID-19 pandemic".

In January 2021, the shipping industry issued the Neptune Declaration on Seafarer Wellbeing and Crew Change, which by June 2021 had been signed by more than 600 companies and organizations. They have also produced a Neptune Declaration Crew Change indicator which aggregates data from 10 leading ship managers which collectively have about 90,000 seafarers currently on board. This reported that between June and July 2021 the situation appeared to be worsening, with more seafarers on vessels beyond the expiry of their contract and more who had been on board for over 11 months – the maximum length of time envisaged in the 2006 Maritime Labour Convention (MLC). Since the launch of the indicator in May 2021, the proportion of seafarers on vessels beyond the expiry of their contract had risen from 5.8 to 8.8 per cent while the proportion on board for over 11 months had increased from 0.4 to 1.0 per cent.

Advances in international law and technology

The COVID-19 pandemic has interfered with international trade, creating inefficiencies, delays and supply-chain disruptions on an unprecedented scale – which also have legal consequences if contractual performance is disrupted, delayed, or becomes impossible. For shipping this can lead to litigation that raises complex international jurisdictional issues. Government and industry will need to work together to address the related contractual rights and obligations, and arrive at standard contractual clauses for commercial risk-allocation.

Many of the problems are associated with delays in documentation – which should encourage more commercial parties to adopt secure electronic solutions. Updated industry guidelines adopted recently, offer useful guidance to shipowners and operators on procedures and actions to maintain the security of IT systems in their companies and onboard ships, adopting a cyber-risk management approach, and taking account of the IMO requirements, and other relevant guidelines.

Technological innovation is also raising the prospect of automated crewless vessels. The industry is conducting trials on "maritime autonomous surface ships" (MASS). The aim is to ensure safe, secure and environmentally sustainable shipping with the relevant legal framework. In May 2021, the IMO Maritime Safety Committee completed a regulatory scoping exercise for the use of MASS which highlighted some priority issues. The outcome could be a MASS instrument/code, with goals, functional requirements and corresponding regulations, suitable for different degrees of autonomy.

On the path to a 3°C temperature rise

The shipping industry has an important part to play in combatting climate change. The Paris Agreement aimed to reduce global warming to well below 2°C and pursue 1.5°C. But, despite a brief dip in carbon dioxide emissions caused by the COVID-19 pandemic, the world is still heading for a temperature rise in excess of 3°C this century. Urgent action is needed on both mitigation and adaptation.

At the regulatory level, the shipping industry is addressing climate issues through the 1973/1978 International Convention for the Prevention of Pollution from Ships (MARPOL). In June 2021, the IMO adopted amendments to Annex VI of the Convention, which introduced new mandatory regulations to further reduce greenhouse gas emissions from shipping, and require owners to set energy efficiency targets. There were also initial discussions on the mid- and long-term action needed, including market-based measures, along with an industry-led proposal for an International Maritime Research and Development Board a non-governmental body which would be financed by a levy on marine fuel and would support research, development, and the deployment of zero-carbon technologies.

Climate change, with the prospect of accelerating sea-level rise and more extreme weather events, will also have major implications for the world's seaports. Securing global maritime transport and trade will therefore mean investing in adaptation and building resilience- for seaports and other key transport infrastructure, especially in developing countries.

Broad-based global recovery will depend on smart, resilient and sustainable maritime transport

The COVID-19 pandemic triggered a succession of shocks and waves, each setting off their own spinoff events. The extent and impact of disruption varied considerably, however, between regions, economic sectors, and segments of the shipping market. The recovery is similarly proving uneven, with differences in the levels and scale of policy support and unequal access to vaccines.

Although the initial impact on maritime transport was less dramatic than predicted, the outlook is shadier. The timescale for a lasting recovery will depend on the progress of the pandemic, the extent and timing of world vaccination plans, and the duration of policy support measures. At present the nascent recovery is being threatened by supply-chain breaks and logistical bottlenecks that are disrupting shipping markets and pushing cost levels to historic highs.

The COVID-19 disruption has also accelerated pre-existing megatrends – geopolitical, technological, and environmental. These trends have been unfolding slowly over the past decade but have accelerated during the pandemic and continue to transform maritime transport and trade:

Geopolitics – The COVID-19 health crisis underscored the extent to which nations are economically and socially interdependent – integrated through global supply chains and their underlying extended maritime transport networks. In the face of heightened geopolitical risks and rising trade tensions, many countries and enterprises are shifting their mindsets and now perceive global interdependency partly as a vulnerability. To mitigate risks and build resilience – they are therefore aiming to reduce their reliance on distant foreign suppliers.

Resilience – The COVID-19 disruption has tested supply chains and their underlying business models, and put transport and logistics networks under strain. Enterprises and governments are aiming to make supply chains more robust and resilient, including by looking to diversify their business partners and suppliers. This will involve a new balance between local, regional and global production. They are also reconsidering inventory and stock management strategies and the trade-offs between just-in-time and just-in-case supply chain models.

Technology – Customs officials, port workers, and transport operators increasingly recognize the value of new technologies and digitalization, not just as a way of boosting efficiency but also for maintaining business continuity at times of disruption. Technological innovations include advanced analytics, on-board sensors, communications technology, port-call optimization, blockchains, big data, and autonomous ships and vehicles. During the pandemic, these technologies have helped reduce physical contact, and keep ships moving, ports open and cross-border trade flowing. Technological advances have also stimulated consumer spending online and a growth in e-commerce. These trends will continue to redefine production and consumption patterns and the ways in which ships, ports and their hinterland connections deliver cargo and services.

Shipping market dynamics – In anticipation of future disruptions, carriers, shippers, ports, and inland transport operators will be rethinking their business and operating models to respond more flexibly to changing market conditions. Having seen the way in which the trade rebound stumbled against logistical bottlenecks and constrained capacity following the COVID-19 shock, they are likely to reconsider their levels of investment in shipping and ports as well as their planning operations. They can also anticipate potential greater regulation of shipping markets as national competition authorities step up their monitoring of freight rates and market behaviour and scrutinize rapid movements in shipping prices.

Decarbonization and the energy transition – Maritime transport is facing growing pressure to decarbonize and operate in a more sustainable way – issues that have also come to the fore as part of the post-pandemic recovery. With ongoing IMO work on greenhouse gas emission reduction in shipping providing further momentum, shipping is expected to change its fuel mix and use new technology and ship designs, alternative fuels and operational adjustments to cut its carbon and environmental footprint. For energy, shipping is not just a large-scale user but also a major carrier, so the industry will have to respond to lower demand for oil tankers and coal carriers and more for ships transporting hydrogen, ammonia and other alternative fuels.

Climate adaptation and resilience – Maritime transport infrastructure and services came under severe stress as a result of the pandemic and the closure of the Suez Canal. This was in addition to the ongoing dangers of climate change: over recent years extreme weather events, including floods, hurricanes and cyclones, have been causing frequent and intense disruptions for both coastal infrastructure and hinterland connections. With current climate projections pointing to a global warming trajectory exceeding the agreed targets under the Paris Agreement, the maritime industry and governments need to invest in adaptation and in climate-proofing maritime transport infrastructure and services, as well as accelerate the development of related legal, policy and technical measures, and capacity-building.

Priorities for action

1. *Vaccinate the world* – To complete broad-based global vaccination, developing countries should have fair access to vaccines. Investing in global vaccination, with the support of dedicated funds, will not just accelerate the end of the pandemic but also stimulate the recovery and add trillions to global economic output.

2. *Revitalize the multilateral trade system* – Decades of trade liberalization and multilateral action have brought economic and social benefits that are now under threat from increasing trade restrictions and protectionism. To retain these hard-won gains countries will need to defend and consolidate the multilateral trade system and minimize trade restrictiveness.

3. *End the crew-change crisis* – This requires urgent attention from flag, port and labour-supplying states, in collaboration with relevant international organizations. All states should be parties to the relevant international legal instruments, including the MLC 2006, ILO Conventions Nos. 108 and 185 on Seafarers' Identity Documents, and the IMO FAL Convention. To advance the objectives of SDG 8, and to ensure decent work for seafarers, states also need to redouble their efforts to ensure that these conventions and labour standards are fully implemented.

4. *Vaccinate seafarers* – Concerted collaborative efforts by industry, governments and international organizations should ensure that seafarers are designated as key workers and are vaccinated as a matter of priority.

5. *Facilitate crew changes* – Governments and industry should continue to work together, including through the Neptune Declaration initiative, and in collaboration with relevant international organizations, to facilitate crew changes, in accordance with international standards and in line with public health considerations. They should also ensure the availability and access to related seafarer data.

6. *Ensure reliable and efficient maritime transport* – Stakeholders in the maritime supply chain, including carriers, ports, inland transport providers and shippers, should work together to ensure that maritime transport remains a reliable, predictable and efficient mode of transport. This will require investing in shipping and ports and their hinterland connections while devising and implementing sustainable freight transport solutions. It will also require proper implementation of trade facilitation measures and digital tools and technologies.

7. *Mainstream supply chain resilience, risk assessment and preparedness* – This can be achieved through a portfolio of measures, including dual sourcing, redundancy across suppliers, and backing up production sites, inventory, and stocks, along with better risk management, and end-to-end transparency. Typically, this will involve assessing and managing risks, enhancing preparedness and adopting hybrid solutions that are flexible and agile, and arrive at balanced trade-offs, for example, between nearshoring and reshoring and combining hybrid supply chain models, along with measures to reduce vulnerabilities to cyberattacks.

8. *Control costs* – Freight costs can be contained by expanding capacity to match demand, making ports more efficient, improving planning, forecasting and visibility, and implementing trade facilitation measures. The maritime transport market should also be transparent, fair and competitive. National competition authorities therefore need the capacity to monitor trends in freight rates, fees and charges. Stakeholders along the maritime supply chain including carriers, ports, inland transport providers, customs, and shippers should work together to share information and make maritime transport more efficient.

9. *Decarbonize* – The shipping industry, in cooperation with governments, will need to explore alternative fuels, invest in landside infrastructure and replace older vessels with larger and more fuel-efficient ships. This will require a predictable environment at the global level but in addition, structurally weak developing countries will need help to mitigate transition costs and the lower connectivity that could result from decarbonizing maritime transport. Developing countries will also need to gain a better understanding of how new regulations will affect the maritime transport services. Integrated post-pandemic recovery planning and stimulus packages should earmark resources for environmental sustainability, aiming for green, low-carbon maritime transport.

10. *Climate-proof maritime transport* – Countries should anticipate, prepare for and adapt to climate change by fully understanding the risks, exposure, and vulnerabilities, and by building adaptive capacity across the maritime supply chain. For developing countries, including the most vulnerable groups of countries, building back better after the pandemic will mean scaling up investment and building national capacities in climate-proofing.

In 2020, international maritime trade and global supply chains were hit by the impact of the COVID-19 pandemic. Overall however, maritime transport managed to navigate through the crisis, and for some parts of the supply chain the impact was not as dramatic as initially feared. Carriers were able to mitigate the early shock and manage lower levels of demand. Port and landside operations, however, struggled to adjust, and the world's seafarers faced a precarious situation as they became caught up in an unprecedented global crew-change crisis.

In 2020, global economic output fell by 3.5 per cent and merchandise trade by 5.4 per cent, while international maritime shipments fell by 3.8 per cent, to 10.65 billion tons. However, UNCTAD expects world maritime trade to recover by 4.3 per cent in 2021, and growth is projected to continue over the 2022–2026 period, albeit at rates that will be moderated by the easing in world economic output. Although the short-term outlook is positive, the medium- and longer-term prospects remain uncertain: the upturn will be directed by the future path of the pandemic and the associated lockdowns and restrictions. A lasting recovery also hinges on keeping trade flowing, by creating supportive macroeconomic and fiscal conditions while minimizing trade protectionism.

Throughout 2021, much of the global economic revival will be driven by government spending in major economies, so the patterns and geography of the recovery will be shaped by the ways in which their governments wind up these support measures – in terms of scale, focus, and timing. Progress could, however, still be derailed by further outbreaks of the pandemic, by slow vaccine deployment and in many economies by the limited scope for policy support. It has become clear that broad-based recovery will require an end to the health crisis and an equitable distribution of vaccines across all regions, developed and developing.

Starting in late 2020, a swift rebound in containerized trade stumbled against supply-side constraints – which increased costs, dented reliability of service, and undermined the operation of value chains. As global demand patterns normalize, these problems are likely to dissipate, but the longer- term outlook will continue to be shaped by wide-ranging and longer-term structural factors, including patterns of globalization, changes in consumption habits, digitalization and the growth of ecommerce, as well as by the global energy transition and the imperative of environmental sustainability.

The impact of COVID-19 has also highlighted the need for better risk management, and greater preparedness, and resilience. The disruption was amplified by other events that created transport bottlenecks – in some countries by flooding, for example, and especially by the blocking of the Suez Canal, which exposed risks and vulnerabilities in supply chains. Building future resilience will entail reforming business models and global supply chains, and reorganizing maritime transport networks.

This chapter considers developments in maritime transport and trade during 2020 until mid-2021. Section A reviews the situation of international maritime trade and container port traffic. Section B sets out the outlook for global recovery and its sustainability. Section C puts forward some key policy considerations and action areas.

International maritime trade and port traffic

Maritime trade and port cargo traffic

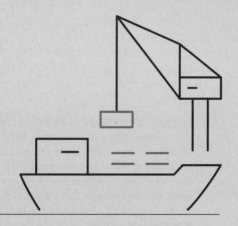

INTERNATIONAL SEABORNE TRADE IN 2020

Growth slipped by
-3.8%
following on a weak pre-pandemic growth of 0.5% in 2019

Total volumes reached
10.7 billion tons

Developing countries continue to account for the lion's share of world maritime trade by volume

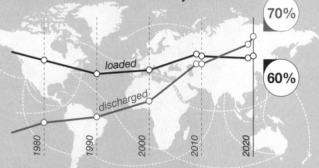

loaded 70%
discharged 60%

World maritime trade, percentage share per region

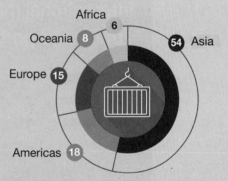

- Asia 54
- Americas 18
- Europe 15
- Oceania 8
- Africa 6

WORLD CONTAINER PORT TRAFFIC IN 2020

Down by
-1.2%

Volumes estimated at
815.6 million TEUs

2020 and 2021 exposed ports' vulnerability to disruptions and risks

World container port traffic by region, 2019-2020 (percentage annual change)

Asia	Africa	Latin America and the Caribbean
-0.4	0.0	-1.8

Europe	North America	Oceania
-4.2	-1.9	-0.8

OUTLOOK

Short-term outlook for maritime trade is positive, however, risks are manifold and uncertainty remains

- Covid-19 pandemic
- Uncertainty
- Congestion in ports

- Supply chain disruption
- Change in globalization patterns
- Transport costs

UNCTAD expects world maritime trade to recover by
+4.3% in 2021

Growth in maritime trade volumes expected to moderate and expand at an annual rate of
+2.4%
between 2022 and 2026

A. VOLUMES OF INTERNATIONAL MARITIME TRADE AND PORT TRAFFIC

The demand for maritime transport services and infrastructure can be assessed through key indicators on trade and port cargo handling. Over the review period, these followed a rollercoaster ride: in early 2020 demand tanked as a result of the pandemic but then bounced back in the second half.

1. International maritime trade fell in 2020 as the pandemic sequentially disrupted supply, demand, and logistics

In 2020, the pandemic disrupted the world economy, cutting manufacturing activity and consumption – with impacts on supply, demand and logistics. International maritime trade growth had already been weak in 2019 at 0.5 per cent, but in 2020 it declined by 3.8 per cent. Total volume dropped by 422 million to 10.65 billion tons (table 1.1 and table 1.2).

Nevertheless, the impact was not as dramatic as initially feared and the maritime transport sector managed to navigate through the crisis (figure 1.1). In 2020, maritime trade increased as a proportion of global GDP, with an increase in the maritime trade-to-GDP ratio as the pandemic induced a shift in consumer demand from services to traded goods. However, this is likely to be short lived as demand patterns normalize and spending continues to rebalance back towards services. In 2021, the narrative is still being driven by the pandemic and related risks, but attention is now moving toward the vaccine rollout, the recovery in growth, and the supply and demand pressures that are currently disrupting trade logistics. At the same time, the industry must consider the longer-term sustainability and resilience of shipping, ports and their hinterland connections.

Around two-thirds of global trade in goods takes place in developing countries (figure 1.2). As indicated in table 1.2, in 2020, developing countries, including the transition economies of Asia, accounted for 60 per cent of global goods loaded (exports) and 70 per cent of goods discharged (imports). Much of this growth has been in East Asia, especially China, and there has also been a surge in volumes on the Transpacific containerized trade route linking East Asia to North America. A smaller proportion of trade was in developed countries, which generated 40 per cent of global maritime exports (goods loaded) and 31 per cent of imports (goods discharged).

Asia's predominance was further strengthened in 2020 as it maintained its 41 per cent contribution to total goods loaded and increased its contribution to total goods discharged (table 1.2 and figure 1.3). Developing America and Africa maintained their existing, smaller shares.

Table 1.1 International maritime trade, 1970–2020
(millions of tons loaded)

Year	Tanker trade[a]	Main bulk[b]	Other dry cargo[c]	Total (all cargoes)
1970	1 440	448	717	2 605
1980	1 871	608	1 225	3 704
1990	1 755	988	1 265	4 008
2000	2 163	1 186	2 635	5 984
2005	2 422	1 579	3 108	7 109
2006	2 698	1 676	3 328	7 702
2007	2 747	1 811	3 478	8 036
2008	2 742	1 911	3 578	8 231
2009	2 641	1 998	3 218	7 857
2010	2 752	2 232	3 423	8 408
2011	2 785	2 364	3 626	8 775
2012	2 840	2 564	3 791	9 195
2013	2 828	2 734	3 951	9 513
2014	2 825	2 964	4 054	9 842
2015	2 932	2 930	4 161	10 023
2016	3 058	3 009	4 228	10 295
2017	3 146	3 151	4 419	10 716
2018	3 201	3 215	4 603	11 019
2019	3 163	3 218	4 690	11 071
2020	2 918	3 181	4 549	10 648

Sources: Compiled by the UNCTAD secretariat based on data supplied by reporting countries and as published on the relevant government and port industry websites, and by specialist sources. Dry cargo data for 2006 onwards has been revised and updated to reflect improved reporting, including more recent figures and a better breakdown by cargo type. Since 2006, the breakdown of dry cargo into "Main bulk" and "Other dry cargo" is based on various issues of the *Shipping Review and Outlook* and *Seaborne Trade Monitor*, produced by Clarksons Research. Total maritime trade figures for 2020 are estimated based on preliminary data or on the last year for which data were available.

[a] Tanker trade includes crude oil, refined petroleum products, gas, and chemicals.

[b] Main bulk includes iron ore, grain, coal, bauxite/alumina, and phosphate. Starting in 2006, "Main bulk" includes iron ore, grain, and coal only. Data relating to bauxite/alumina and phosphate are included under "Other dry cargo".

[c] Includes minor bulk commodities, containerized trade, and residual general cargo.

Table 1.2 International maritime trade 2019–2020, by type of cargo, country group and region

		Goods loaded				Goods discharged			
	Year	Total	Crude oil	Other tanker trade[a]	Dry cargo	Total	Crude oil	Other tanker trade[a]	Dry cargo
		Millions of tons							
World	2019	11 070.5	1 860.3	1 302.6	7 907.6	11 055.1	2 022.8	1 320.5	7 711.8
	2020	10 648.3	1 716.0	1 202.3	7 730.0	10 631.1	1 863.6	1 222.0	7 545.5
Developed economies	2019	4 503.2	453.6	477.1	3 572.6	3 778.3	902.0	463.3	2 412.9
	2020	4 317.4	425.9	430.3	3 461.2	3 245.2	732.5	370.2	2 142.5
Developing economies	2019	6 567.3	1 406.7	825.5	4 335.1	7 276.8	1 120.7	857.2	5 298.9
	2020	6 330.9	1 290.1	772.0	4 268.8	7 385.9	1 131.2	851.7	5 403.0
Africa	2019	814.1	302.8	91.6	419.6	533.7	35.3	113.4	385.0
	2020	735.5	236.1	83.4	415.9	510.1	30.6	107.9	371.5
Latin America and the Caribbean	2019	1 406.6	221.9	81.3	1 103.3	621.4	45.0	143.7	432.6
	2020	1 369.2	200.5	75.6	1 093.1	590.1	39.6	130.0	420.5
Asia	2019	4 331.4	880.1	644.6	2 806.6	6 108.0	1 039.6	595.6	4 472.7
	2020	4 212.2	851.8	605.8	2 754.5	6 272.4	1 060.2	609.6	4 602.6
Oceania	2019	14.5	1.7	7.8	5.0	14.9	0.8	5.4	8.6
	2020	14.6	1.8	7.8	5.1	15.4	0.7	5.5	9.1

		Goods loaded				Goods discharged			
	Year	Total	Crude oil	Other tanker trade[a]	Dry cargo	Total	Crude oil	Other tanker trade[a]	Dry cargo
		Percentage share							
World	2019	100.0	16.8	11.8	71.4	100.0	18.3	11.9	69.8
	2020	100.0	16.1	11.3	72.6	100.0	17.5	11.5	71.0
Developed economies	2019	40.7	24.4	36.6	45.2	34.2	44.6	35.1	31.3
	2020	40.5	24.8	35.8	44.8	30.5	39.3	30.3	28.4
Developing economies	2019	59.3	75.6	63.4	54.8	65.8	55.4	64.9	68.7
	2020	59.5	75.2	64.2	55.2	69.5	60.7	69.7	71.6
Africa	2019	12.4	21.5	11.1	9.7	7.3	3.2	13.2	7.3
	2020	11.6	18.3	10.8	9.7	6.9	2.7	12.7	6.9
Latin America and the Caribbean	2019	21.4	15.8	9.8	25.5	8.5	4.0	16.8	8.2
	2020	21.6	15.5	9.8	25.6	8.0	3.5	15.3	7.8
Asia	2019	66.0	62.6	78.1	64.7	83.9	92.8	69.5	84.4
	2020	66.5	66.0	78.5	64.5	84.9	93.7	71.6	85.2
Oceania	2019	0.2	0.1	1.0	0.1	0.2	0.1	0.5	0.2
	2020	0.2	0.1	0.9	0.1	0.2	0.1	0.5	0.2

Source: Compiled by the UNCTAD secretariat based on data supplied by reporting countries and as published on the relevant government and port industry websites, and by specialist sources. Dry cargo data for 2006 onwards has been revised and updated to reflect improved reporting, including more recent figures and a better breakdown by cargo type. Total maritime trade figures for 2020 are estimated based on preliminary data or on the last year for which data were available.

Note: Since March 2021, the category "transition economies" is no longer used by UNCTAD. Economies formerly classified as "transition economies" and located in Europe, are reassigned to the "developed regions" grouping, and the economies formerly classified as "transition economies" and found in Asia, are reassigned to the "developing regions" grouping. For more extended time series and data before 2020 see UNCTADstat Data Center at https://unctadstat.unctad.org/wds/TableViewer/tableView.aspx?ReportId=32363. Annual world totals of goods loaded and discharged are not necessarily the same, given among other factors, bilateral asymmetries in international merchandise trade statistics and the fact that volumes loaded in one calendar year may reach their port of destination in the next calendar year.

[a] Include crude oil, refined petroleum products, gas, and chemicals.

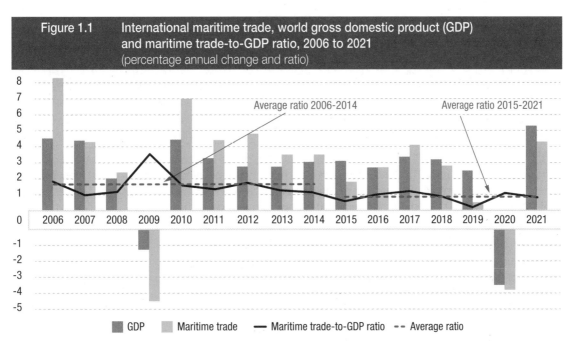

Figure 1.1 International maritime trade, world gross domestic product (GDP) and maritime trade-to-GDP ratio, 2006 to 2021
(percentage annual change and ratio)

Source: UNCTAD calculations, based on the *Review of Maritime Transport,* various issues, data from UNCTADstat and table 1.1 of the UNCTAD *Trade and Development Report 2021. From Recovery to Resilience: The Development Dimension.*

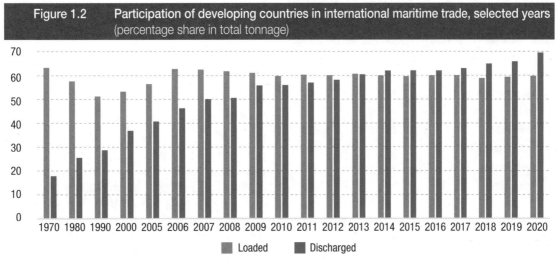

Figure 1.2 Participation of developing countries in international maritime trade, selected years
(percentage share in total tonnage)

Source: UNCTAD secretariat based on the *Review of Maritime Transport*, various issues, and table 1.2 of this report.

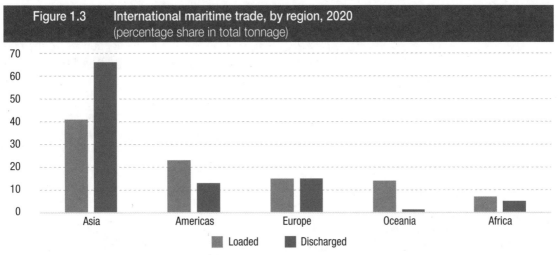

Figure 1.3 International maritime trade, by region, 2020
(percentage share in total tonnage)

Source: Compiled by the UNCTAD secretariat based on data supplied by reporting countries and as published on the relevant government and port industry websites and by specialist sources.

2. Disruption of global economy and trade followed by signs of a multi-paced recovery

In 2020, global GDP declined by 3.5 per cent (table 1.3) – the largest downturn for 70 years. The greatest impact was in the services sector – in particular in tourism, travel and hospitality. For maritime trade, however, the plunge in flows was mitigated by the boost in demand from government stimulus packages. Estimated in March 2021 at around $16 trillion, and concentrated mainly in the United States, Europe and Japan, these packages helped soften the landing. Demand has further revived with the lifting of some COVID-19-related restrictions.

By the third quarter of 2020, there were signs of recovery, driven by positive trends in East Asia and the United States and the rollout of COVID-19 vaccines in many developed economies. While the manufacturing sector was down, consumer demand rose, notably in the United States with end-year retail sales 3.4 per cent higher than 2019 (Sand, 2021a). Unlike the downturn in the first half of 2020, however, which was globally synchronized, the nascent recovery is proceeding along diverging tracks, as many other economies, especially in developing regions continue to fall behind.

In 2020 the drop in GDP in developing economies, at 1.8 per cent (table 1.3), was less than the global average of 2.9 per cent for the 2009–2021 period. This was largely due to the performance of China which was the only country to have seen some economic growth in 2020 (2.3 per cent). China's efforts to contain the pandemic, along with a stimulus package, provided support to industry and exports.

In 2020, output in developed economies contracted by 4.7 per cent. The drop was lower in the United States at 3.5 per cent, as fiscal measures helped minimize the economic downturn, and steeper in the EU at 6.2 per cent, reflecting renewed pandemic outbreaks. In the United Kingdom, the drop was steeper still at 9.9 per cent, as a result not just of the pandemic restrictions but also of Brexit which disrupted supply chains as traders adjusted to new rules and procedures. Elsewhere, Japan's economy fell by 4.7 per cent while India's dipped by 7.0 per cent. There was also a severe impact on GDP in Latin America and the Caribbean, down by 7.1 per cent, in Africa by 3.4 per cent, in Western Asia by 2.9 per cent, and the Russian Federation by 3.0 per cent.

For 2021, current projections for global GDP are pointing to growth of 5.3 per cent. Progress is again expected to be uneven, with Asia and the United States forging ahead. The speed and geography of the recovery will depend to large extent on the vaccine rollout and on the structure, scale, and duration of government support, as for example, in:

- *India* – The announced support measures focus on road infrastructure and are expected to boost dry bulk shipping by increasing demand for raw materials.

- *Japan* – The $3-trillion stimulus package, including the funds announced at the end of 2020 and focusing on green and digital innovation, could boost container volume in intra-Asian trade.

- *United States* – Additional fiscal stimulus measures, including large infrastructure plans will lift demand for some commodities.

- *European Union* – Spending from the Next Generation recovery fund is due to begin in 2021.

- *Least developed countries* – Stimulus packages average only 2.1 per cent of their GDP, i.e., one-ninth of the global average (UNDESA, 2021).

Table 1.3 World economic growth, 2019–2021 (annual percentage change)

Region or country	2019	2020	2021[a]
World	2.5	-3.5	5.3
Developed countries	1.7	-4.7	4.7
of which:			
United States	2.2	-3.5	5.7
European Union (27)	1.6	-6.2	4.0
United Kingdom	1.4	-9.9	6.7
Japan	0.3	-4.7	2.4
Australia	1.8	-2.5	3.2
Russian Federation	1.3	-3.0	3.8
Developing countries	3.7	-1.8	6.2
of which:			
Africa	2.9	-3.4	3.2
East Asia	4.3	0.3	6.7
of which:			
China	6.1	2.3	8.3
South Asia	3.1	-5.6	5.8
of which:			
India	4.6	-7.0	7.2
South-East Asia	4.4	-3.9	3.5
Western Asia	1.3	-2.9	3.5
Latin American and the Caribbean	-0.1	-7.1	5.5
of which:			
Brazil	1.4	-4.1	4.9

Source: UNCTAD secretariat, based table 1.1 of UNCTAD Trade and Development Report 2021. From Recovery to Resilience: The Development Dimension.

Note: Calculations for country aggregates are based on world GDP at constant 2015 dollars.

[a] Forecast.

Table 1.4 Growth in the volume of world merchandise trade, 2019–2021
(annual percentage change)

	Volume of exports (percentage change)			Volume of imports (percentage change)		
	2019	2020	2021[a]	2019	2020	2021[a]
World	**-0.3**	**-5.3**	**14.3**	**-0.3**	**-5.5**	**13.3**
Developed countries	**-0.2**	**-6.7**	**12.5**	**-0.2**	**-5.6**	**12.2**
of which:						
Euro area	-0.1	-8.7	13.4	0.0	-8.2	11.3
United States	-0.5	-11.0	11.0	-0.4	-4.0	16.0
United Kingdom	-3.1	-14.4	-2.5	3.9	-13.5	7.7
Japan	-1.6	-7.8	17.3	0.8	-6.2	3.7
Other developed countries	**2.0**	**-5.1**	**12.3**	**0.0**	**-4.5**	**15.3**
Developing countries	**-0.4**	**-2.3**	**17.5**	**-0.6**	**-5.2**	**15.9**
of which:						
China	0.4	1.3	34.3	0.0	1.7	17.1
Latin America	0.6	-4.2	9.9	-1.5	-11.2	21.0
Africa and the Middle East	-4.0	-6.8	-2.7	-0.3	-2.8	3.1
Asia (not including China)	-1.3	-3.6	19.6	-2.4	-11.6	20.2
Eastern Europe and Commonwealth of Independent States	**2.0**	**-2.2**	**0.6**	**5.0**	**-5.4**	**8.8**

Source: UNCTAD Secretariat calculations, based on CPB World Trade Monitor, July 2021. Data source and methodology are aligned with UNCTAD, *Trade and Development Report 2021.*

Note: Country coverage and classification in the aggregated country groupings is not comprehensive and relies on Ebregt (2020).

[a] For 2021, figures reflect percentage change between the average for the period January to May 2021 and January to May 2020.

In 2020 taken together, world merchandise imports and exports fell by 5.4 per cent (table 1.4), This decline was far lower than more pessimistic forecasts at the height of the pandemic (UNCTAD, 2020a). In April 2020, the World Trade Organization (WTO) had expected world merchandise trade to drop by between 13 and 32 per cent in 2020 (WTO, 2020). There was indeed a slump in the second quarter of 2020 but trade volumes bounced back in the third quarter, responding to the easing of restrictions and lockdowns and announcements of new vaccines. Along with vaccine rollout in major developed regions, the rapid return in volumes reflected the resilience of East Asian trade and the boost in consumer demand from fiscal spending in the United States. Trade in services however remained subdued across all economies. Tourism and cruise shipping were hit hard, though there was a growth in cross-border services that were increasingly enabled by digital technologies.

Exports and imports fell in almost all regions – though to different extents. As shown in table 1.4, between 2019 and 2020 developed country regions saw a drop in exports of 6.7 per cent and in imports of 5.6 per cent. The United Kingdom recorded a double-digit drop in exports, as did the United States though here the implementation of the Phase One trade agreement boosted some exports to China (Sand, 2020a). Trade also declined in the euro area and Japan albeit at relatively lower rates while trade involving other developed regions fared relatively better with exports falling by only 5.1 per cent and imports by 4.5 per cent.

Developing regions also recorded a drop in merchandise trade volumes although at more moderate rates: exports fell by 2.3 per cent while imports dropped by 5.2 per cent. The one exception was China where, despite the disruption, exports rose by 1.3 per cent and imports by 1.7 per cent. For Asia, excluding China, however, exports declined by 3.6 per cent while imports dropped by 11.6 per cent. In Latin America imports dropped by 11.2 per cent and exports by 4.2 per cent. In Africa and the Middle East exports fell by 6.8 per cent and imports by 2.8 per cent. In Eastern Europe and Commonwealth of Independent States, the decline in imports was less at 2.2 per cent, though imports fell by 5.4 per cent.

2021 saw a revival in world merchandise trade. During the first five months of the year exports were 14.3 per cent higher than in the corresponding period in the previous year, while imports rose by 13.3 per cent (table 1.4). But the recovery was uneven with exports from Africa and the Middle East as well as from the United Kingdom continuing their decline. In the United States imports jumped by 16.0 per cent, reflecting inventory building and the lasting benefits of fiscal support measures. During the same period, imports increased into the euro area by 11.3 per cent, the United Kingdom by 7.7 per cent and Japan by 3.7 per cent. Imports into developing countries increased by 15.9 per cent and into Eastern Europe and Commonwealth of Independent States by 8.8 per cent.

Much of global import demand in the first half of 2021 was met from Asia, in particular from China whose exports expanded by 34.3 per cent. There was also stronger import growth in Latin America, of 21.0 per cent. Recovery in Africa and the Middle East was more moderate for both exports and imports. For the full year 2021, the WTO expects world merchandise trade volume to grow by 8.0 per cent though the recovery will be uneven (WTO, 2021).

This bounce-back in merchandise trade in almost all major economies has been faster than in previous recessions – in 2009 and 2015 – though it has been from a low base and has been more robust in goods than services (UNCTAD, 2021). The rebound was evident across a wide range of sectors including pharmaceuticals, communications and office equipment, as well as minerals and agri-food. Much of this has been due to the release of pent-up demand for durable goods such as cars, as well as strong demand for products that support working from home. In contrast, recovery in the energy sector remains hesitant.

3. Maritime trade fell in 2020 but fared better than initially feared

The sudden dip and subsequent recovery in merchandise trade was reflected in the patterns of maritime trade. In 2020, the outcome was better than initially feared. Volumes dipped by around 12 per cent in May 2020 compared with May 2019, but only by around 2.0 per cent in the fourth quarter compared with the same quarter in 2019 (Clarksons Research, 2021b). For 2020, following a contraction of 3.8 cent, UNCTAD estimates shipping volumes to have lost 422 million tons.

The performance varied by market segment, with some sectors performing better than others (table 1.1, table 1.2, figure 1.4). Worst hit was tanker shipping, but there was less impact on containerized trade, gas shipments, and on dry bulk commodities such as iron ore and grains.

The second half of 2020 saw a nascent recovery – though asymmetric across market segments. There was a return in volumes for containerized and dry bulk commodities, but tanker shipping awaited a full recovery in global demand. At the same time, the sudden boost in demand stumbled into shortages – of shipping capacity, and of containers, and equipment. As result, freight rates surged, with proliferating surcharges. This may have bolstered shipping profitability but it put supply chains under strain, while adding to port congestion and increasing delays and dwell times, and leading to a general decline in service reliability.

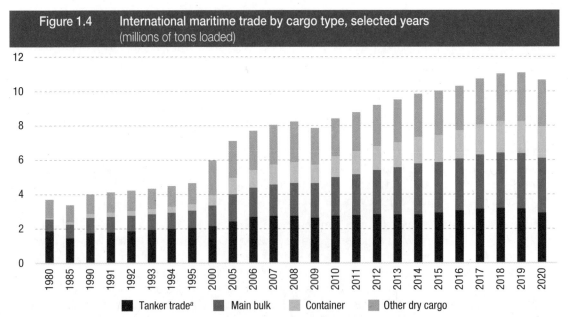

Figure 1.4 International maritime trade by cargo type, selected years (millions of tons loaded)

Source: UNCTAD Review of Maritime Transport, various issues. For 2006–2020, the breakdown by cargo type is based on Clarksons Research, Shipping Review and Outlook, Spring 2021 and Seaborne Trade Monitor, various issues.

Note: Given methodological differences, containerized trade data in tons sourced from Clarksons Research are not comparable with data in TEUs featured in tables 1.8 and 1.9 and figures 1.8 and 1.9 of this report and which are sourced from MDS Transmodal.

[a] Tanker trade includes crude oil, refined petroleum products, gas, and chemicals.

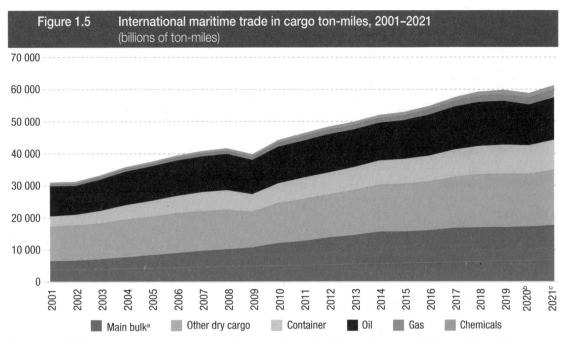

Source: UNCTAD secretariat based on data from Clarksons Research. *Shipping Review and Outlook*, Spring 2021.
^a *Includes* iron ore, grain, coal, bauxite/alumina, and phosphate.
^b *Estimated.*
^c *Forecast.*

The pandemic has proved to be an asynchronous, multi-wave event, as COVID-19 outbreaks lead to sequences of lockdowns and various restrictions. In 2020 these disruptions were exacerbated by other events such as the closure in China of the port of Yantian, which is a critical international container terminal, and the week-long blockage of the Suez Canal, with further problems in 2021 as a result of extreme weather events. For some of the major industries in Europe, these bottlenecks are causing shortages of inputs and delays in delivery, and generally holding up the recovery. Automotive plants, for example, had to close temporarily due to missing critical components and parts (Ewing and Clark, 2021). This confluence of factors exposed the vulnerabilities of supply chains and of their underlying maritime transport systems. They have also amplified the call for nearshoring and reduced the attractiveness of long-haul trade and extended supply chains.

When adjusted for distance travelled, however, the decline in maritime trade in 2020 was lower – falling by only 1.7 per cent, to an estimated 58,865 billion cargo ton-miles (figure 1.5). But there were different outcomes for different types of cargo: oil decreased by 7.0 per cent and containerized trade by 1.5 per cent, while there was an increase of 1.3 per cent in dry bulk trades (iron ore, coal, and grain) and of 6.7 per cent in gas shipments, including liquified petroleum gas (LPG) and liquified natural gas (LNG) (Clarksons Research, 2021a).

International maritime trade flows were sustained in 2020 by the rapid economic rebound in China with a 9 per cent increase in maritime import demand, in particular imports of iron ore and grain. Maritime trade flows were also supported by China's exports of containerized goods to the United States. Meanwhile, lower demand for oil, and cuts by major OPEC+ oil producers and oil production, have continued to keep a lid on the recovery in tanker shipping.

Most ton-miles and tons generated by bulkers of over 100,000 dwt were contributed by shipments from Australia, followed by Brazil. In 2020, Australia generated 58 per cent of world iron ore exports and Brazil 23 per cent (figure 1.6). Much of this is destined for China. In 2020, China accounted for 76 per cent of world iron ore imports and 20 per cent of coal imports. Tonnage on the Australia-China route, however, declined in 2020, probably as result of the pandemic and the tensions between the two countries. China is seeking to diversify its sources of supply and is looking more to Africa. Trade in ton-miles generated by bulkers on the Africa-China route increased in 2020, probably reflecting increased iron ore shipments from South Africa. Guinea could also be a supplier since it is reported to hold large reserves of untapped high-quality iron ore. Guinea is expected to start shipping iron ore beginning in 2026, which will boost demand for dry bulk shipping (Hellenic Shipping News, 2020). The country is already the world's top supplier of bauxite, much of which is shipped to China.

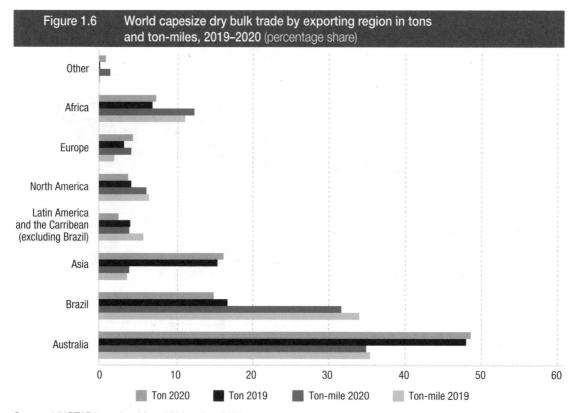

Figure 1.6 World capesize dry bulk trade by exporting region in tons and ton-miles, 2019–2020 (percentage share)

Source: UNCTAD based on VesselsValue data 2021.
Note: Based on dry bulk vessels of more than 100,000 dwt.

Crude oil exports continue to be dominated by Western Asia (figure 1.7). Much of the world's import demand is from Asia, mainly China and India, followed by Japan and the Republic of Korea. Ton-mile increase generated by North American exports in 2020 reflects the strong import demand in China and growth in exports from the United States captured under Phase One of the trade deal with China. At the underlying level, the shale boom is also a key driver of North American oil exports, with the United States becoming a net seaborne energy exporter.

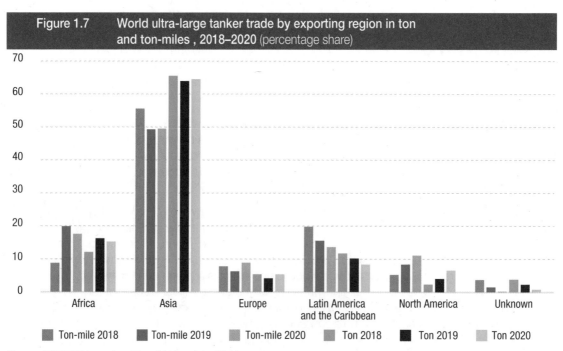

Figure 1.7 World ultra-large tanker trade by exporting region in ton and ton-miles, 2018–2020 (percentage share)

Source: UNCTAD based on VesselsValue data, 2021.
Note: Tanker vessels of more than 320,000 dwt.

4. Diverging impacts and recoveries for key shipping markets

Oil trade still under pressure and gas trade down

The shipping market hardest hit by the pandemic has been the oil trade. Between 2019 and 2020 UNCTAD estimates that tanker trade, including crude oil, refined petroleum products, and gas, slipped by 7.7 per cent, with volumes down from 3.2 billion to 2.9 billion tons (table 1.5).

The steepest drop was for seaborne crude oil at 7.8 per cent, as total volumes fell to 1.7 billion tons. Crude oil imports declined in most key importing markets including the United States, Europe, India, Japan, and the Republic of Korea. The only increase was in China, by 8 per cent.

The demand for crude oil in 2020 reflects a reduction in demand for fuel – Jet A for aircraft, gasoline for automobiles, and diesel for trucks – with volumes declining by over 10 per cent (Clarksons Research, 2021b). While road travel is expected to increase, long-distance aviation prospects remain uncertain, awaiting a worldwide rollout of vaccines.

Fuel imports to West Coast Latin America from the United States have fallen, partly because of limited refinery capacity in the United States, opening up an opportunity for suppliers from Asia. Increased diesel and gasoline shipments from Asia to West Coast Latin America will benefit ton-mile growth (Connelly, 2021).

The tanker trade has suffered from weak oil demand, high inventories, and cuts in oil supply by OPEC+ members. That said, 2021 should see an improvement as demand gradually recovers and supply increases. Starting in August 2021, as oil prices hit their highest levels in more than two years, OPEC+ members agreed to phase out 5.8 million barrels per day of production cuts (OPEC, 2021). Meanwhile, a lifting of the United States sanctions would increase exports from the Islamic Republic of Iran, which could displace production from other locations but nevertheless increase the demand for tankers. With an increase in OPEC production and the expansion of Asian refineries, there is likely to be more demand for very large crude carriers.

India's recent decision to diversify crude oil imports and reduce its dependency on Western Asia is also good news for operators of crude-oil tankers and will boost demand in terms of ton-miles (Drewry Maritime Research, 2021a). Ongoing repositioning of refinery capacity closer to demand is likely to alter trade patterns, which could boost crude ton-miles but is more likely to reduce product tanker ton-miles.

In the longer term, tanker demand will be affected by the current global energy transition, which implies a change in the energy mix. Elsewhere, as more refineries in some advanced economies close, changes to oil trade patterns are likely to intensify (Danish Shipping Finance, 2021). A reduction in the United States exports due to the low oil price environment may reduce long-haul trades. Suezmaxes may regain some business due to the potential expansion of Western Asian crude oil production destined for India and South East Asia (Danish Shipping Finance, 2020). Oil product trade flows could become more regionalised, lowering seaborne volumes and travel distances (Danish Shipping Finance, 2020). Ongoing repositioning of refinery capacity closer to demand is likely to alter trade patterns, which could boost crude ton-miles but would more likely reduce product tanker ton-miles. The pandemic has also weighed, if to a lesser extent, on the global demand for gas. In 2020, global gas trade increased only marginally, by 0.4 per cent, while volumes of LNG exports are estimated to have expanded by 1.1 per cent and of LPG to have declined by 1.0 per cent. Gas projects have been delayed by weak energy prices, including work on LNG export terminals in the United States and LNG feedstock projects in Australia (Clarksons Research, 2020). That said, exports from the United States rebounded in 2020, thanks to a boost in consumer demand supported by a cold winter in Asia. The United States also increased its LPG trade, by 15 per cent.

Natural gas offers a lower-carbon source of energy, so with more demands for sustainability and a transition to lower-carbon energy, the global gas trade is set to increase. Much of the growth will be driven by Asia, with an important role for China's new propane dehydrogenation plants. India's trade will also expand as a result of subsidized domestic LPG prices.

Table 1.5	Tanker[a] trade, 2019–2020 (million tons and percentage annual change)		
	2019	2020	Percentage change 2019–2020
Crude oil	1 860	1 716	-7.8%
Other tanker trade	1 303	1 202	-7.7%
of which:			
Gas	478	480	0.4%
Total tanker trade	**3 163**	**2 918**	**-7.7%**

Sources: UNCTAD secretariat, derived from UNCTAD data in table 1.2 of this report.
Note: Gas trade figures are derived from Clarksons Research, Seaborne Trade Monitor, Volume 8, No.6, June 2021.
[a] Includes: refined petroleum products, gas, and chemicals.

Natural gas is set to contribute a larger share to the global energy mix in the coming years, with much of the growth driven by shale-gas production in the United States, as well as by production in Western Asia and in other regions including the Mediterranean and East Africa (Clarksons Research, 2020).

Dry bulk commodity trade defied pressure in 2020 with China keeping the trade flowing[1]

Total dry bulk trade fell by an estimated 1.5 per cent in 2020, as volumes slipped to 5.2 billion tons (table 1.6). China's rapid economic recovery has boosted its import demand so it could take up extra cargo generated by suppressed demand in other regions. Iron ore trade remained unperturbed as shipments increased by 3.2 per cent to 1.5 billion tons. Grain trade also held firm, increasing volumes by 7.1 per cent. Supporting factors included a record Brazilian harvest, the returning United States-China trade, and better prospects in pig farming in China following the recovery from the 2018 African swine fever outbreak. In 2021, seaborne dry bulk trade is projected to expand by 3.7 per cent, with iron ore and grain trade growing steadily, a rebound in minor bulk volumes and more coal trade.

Coal trade plunged 9.3 per cent in 2020, partly as a result of the pandemic, with reduced electricity demand across regions overlaid on the ongoing structural shift towards cleaner energy sources. Minor bulk trade also came under pressure, though only falling by 2.2. per cent. There was also less trade in forest products, as well as lower nickel ore exports due to Indonesia's export ban. The bauxite trade was much stronger, expanding by 8.2 per cent, with China accounting for 77 per cent, and Guinea providing 46 per cent of the supply (Clarksons Research, 2021b).

The current major players in the dry bulk trade are featured in table 1.7. These patterns are likely to change as a result of tensions between China and Australia which are affecting coal and iron ore trade. To compensate for the ban on Australian cargo China has cut import duties on coal by land from Mongolia. This would reduce trade by ship, though the impact could be mitigated by increases on the Indonesia-China route (Drewry Maritime Research, 2021b). Meanwhile, a shift in Australia's exports away from China to more distant locations such as Saudi Arabia will increase shipping demand and ton-miles (Drewry Maritime Research, 2021c).

Recovering from the pandemic on the 'build back better' principle will require greener and smarter solutions and a shift towards cleaner and lower-carbon energy sources. In the longer term this will undermine demand for dry bulk carriers (Danish Shipping Finance, 2020). Equally, as the Chinese economy becomes less steel intensive, its demand for iron ore will flatten. The loss of seaborne trade could, however, be partially offset by a growth in trade in the non-ferrous metals that are essential for producing renewable technologies – such as nickel ore, copper, lithium, cobalt, and bauxite – though these commodities are mostly traded in smaller volumes (Danish Shipping Finance, 2021).

Trade tensions between China and the United States have affected trade in grain. In 2017, the United States accounted for 34 per cent of China's seaborne grain imports. In 2019, this share fell to 18 per cent, before recovering to 27 per cent in 2020, on the back of the Phase One trade deal commitments. China's efforts to diversify its suppliers have benefited Brazil whose share of the Chinese market increased from 44 per cent in 2017 to about 60 per cent in 2018 and 2019, before falling back to 48 per cent in 2020 (Zhang, 2021). Other countries have also gained market share, including Ukraine, France, the Russian Federation, and Argentina. But China's grain import demand also faces 'downside risks, including a renewed outbreak of African swine fever and softer crush margins that may dampen soybean imports.

Table 1.6	Dry bulk trade 2019–2020 (million tons and percentage change)		
	2019	2020	Percentage change 2019–2020
Main bulk[a]	3 218.0	3 181.0	-1.1%
of which:			
Iron ore	1 456.0	1 503.0	3.2%
Coal	1 284.0	1 165.0	-9.3%
Grain	478.0	512.0	7.1%
Minor bulk	2 030.0	1 986.0	-2.2%
of which:			
Steel products	373.0	354.0	-5.1%
Forest products	383.0	365.0	-4.7%
Total dry bulk	5 248.0	5 167.0	-1.5%

Source: UNCTAD secretariat calculations, based on Clarksons Research, 2019d, *Dry Bulk Trade Outlook*, Volume 26, No. 6, June.

[a] *Includes* iron ore, coal (steam and coking) and grains (wheat, coarse grain and soybean).

---

[1] Detailed figures on dry bulk commodities are derived from Clarksons Research (2021), *Seaborne Trade Monitor*. Volume 8. No. 6. June.

Government fiscal spending boosts consumption and helps containerized trade weather the storm

In 2020, full box trade fell by just 1.1 per cent to 149 million twenty-foot equivalent units (TEU) (figure 1.8). This was a better outcome than initially feared and quite an accomplishment compared to the 8.4 per cent plunge in 2009 following the financial crisis. After the shock in early 2020, volumes swiftly returned, as consumer demand was boosted by stimulus packages and measures to support incomes.

The bounce-back in 2021 reflected easing economic impacts and the unlocking of pent-up demand, as well as restocking and building inventory. But there was also a shift in consumption patterns away from services and towards goods, notably for health products and pharmaceuticals, as well as home office equipment, along with changes in shopping patterns and the expansion of ecommerce. The surge in trade was welcome but on such a scale that shipping services and port operations were often unable to keep up, resulting in logistical bottlenecks. By the end of 2020 and until the first half of 2021, the whole industry, including shipping, ports, shippers, and inland carriers struggled with shortages in containers, equipment and shipping capacity. This has added to port congestion and reduced service levels and reliability, while also increasing freight rates and surcharges (see chapter 3).

Reflecting the rebound in volumes on the eastbound leg of the East Asia-United States trade, the combined share of the East-West trade routes, including the Asia-Europe, the Transpacific, and the Europe-North America (Transatlantic) increased marginally in 2020. Together, intra-regional trade, essentially reflecting Intra-Asian flows and South-South trade, accounted for over 39.5 per cent of the total. Non-mainlane East-West trade routes (e.g., Eastern Asia-South Asia-Western Asia) and North-South routes represented 12.9 per cent and 8.0 per cent of the market, respectively.

Performance varied across regions and trade lanes (table 1.8). In 2020, total volumes on the mainlane routes decreased by only 0.3 per cent, as the declines of 2.6 per cent on the Asia-Europe trade lane and of 3.2 per cent on the Transatlantic lane were partially offset by growth of 2.8 per cent on the Transpacific route (table 1.9). Non-mainlane trade fell by 1.6 per cent, reflecting the disruption in India which reduced the East-West trade by 3.3 per cent. North-South trade fell by 1.8 per cent, while South-South trade contracted by 2.4 per cent. By early summer of 2020 the rapid recovery

Table 1.7 Major dry bulk and steel: producers, users, exporters, and importers, 2020
(percentage share of world markets)

Steel producers		Steel users	
China	56	China	56
India	5	India	6
Japan	4	United States	5
United States	4	Japan	5
Russian Federation	4	Republic of Korea	4
Republic of Korea	4	Russian Federation	4
Turkey	2	Germany	2
Germany	2	Turkey	2
Brazil	2	Viet Nam	1
Islamic Republic of Iran	2	Other	15
Other	15		

Iron ore exporters		Iron ore importers	
Australia	58	China	76
Brazil	23	Japan	7
South Africa	5	Europe	6
Canada	4	Republic of Korea	5
India	3	Other	6
Sweden	1		
Other	6		

Coal exporters		Coal importers	
Indonesia	35	China	20
Australia	31	India	19
Russian Federation	13	Japan	14
United States	5	Republic of Korea	10
South Africa	6	European Union	6
Colombia	5	Taiwan Province of China	6
Canada	2	Malaysia	3
Other	3	Other	22

Grain exporters		Grain importers	
United States	26	East and South Asia	49
Brazil	23	Africa	14
Argentina	11	South and Central America	10
Ukraine	10	Western Asia	9
European Union	9	European Union	9
Russian Federation	7	North America	1
Canada	6	Other	8
Australia	3		
Other	5		

Sources: UNCTAD secretariat, based on data from the World Steel Association (2021), Clarksons Research *Seaborne Trade Monitor*, Volume 8, No. 6, June 2021; *Dry Bulk Trade Outlook*, Volume 27, No.6, June 2021.

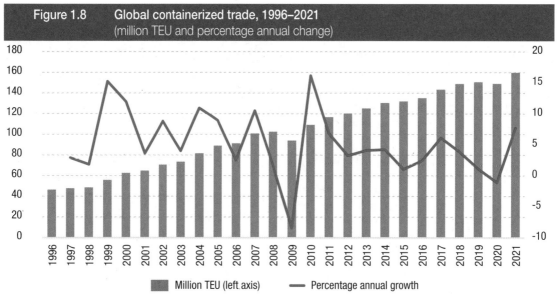

Figure 1.8 Global containerized trade, 1996–2021
(million TEU and percentage annual change)

Source: UNCTAD secretariat calculations, based on MDS Transmodal, World Cargo Database, June 2021.
Note: Projected figure for 2021 based on table 1.11 of this report.

in Asia had helped the intra-Asian trade rebound, and for the full year the decline was only 0.4 per cent for intra-regional trade.

2020 saw an increase of 2.8 per cent on the Transpacific route, boosted by a surge in flows from East Asia to the United States (table 1.9). Between the fourth quarter of 2019 and the first quarter of 2020, containerized trade from Asia to North America had dropped by 13 per cent, but in the third quarter of 2020 it jumped by 36 per cent. While container shipping imports to the United States had been rising, exports from that country had fallen considerably. At the port of Los Angeles, for example, loaded imports were four times greater than loaded exports – so the return legs often had empty containers, which created shortages for exporters.

Faced with congestion and long waiting times at ports, stakeholders have looked for alternatives. In some cases, they have accepted more costly air freight and in others have diverted ships away from the busiest ports. In the short term, these problems are unlikely to diminish. The latest United States $1.9-trillion stimulus package should boost consumer spending which, combined with low inventory levels, is expected to increase imports (Sand, 2021b). In the second quarter of 2021, containerized shipments from East Asia to North America were 35 per cent higher than in equivalent quarter in 2020 (MDS Transmodal, 2021).

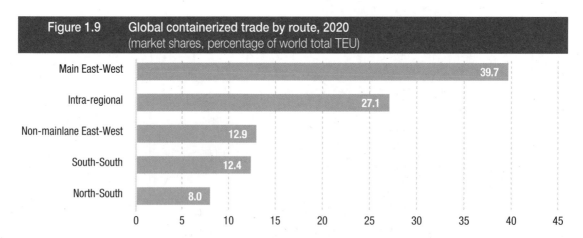

Figure 1.9 Global containerized trade by route, 2020
(market shares, percentage of world total TEU)

Source: UNCTAD secretariat calculations, based on data from MDS Transmodal, World Cargo Database, June 2021.
Note: Non-mainlane East West: Trade involving Western Asia and the Indian Sub-continent, Europe, North America, and East Asia.
North-South: Trade involving Oceania, Sub-Saharan Africa, Latin America, Europe, and North America.
South-South: Trade involving Oceania, Western Asia, East Asia, Sub-Saharan Africa and Latin America.
Intra-regional: Trade within Europe, Africa, Asia, North America, Latin America and Oceania.

Table 1.8 Containerized trade on East-West trade routes, 2016–2020
(million TEU, percentage annual change)

	2016	2017	2018	2019	2020
Main East-West routes	54 480 143	57 520 472	60 323 619	59 317 350	59 168 679
Other routes	80 879 086	86 095 802	88 844 890	91 538 274	90 046 704
of which					
Non-mainlane East-West	18 005 252	19 056 910	19 049 879	19 960 498	19 299 089
North-South	11 120 656	11 745 000	12 086 773	2 099 662	11 882 623
South-South	15 533 787	16 920 644	18 175 418	18 892 469	18 430 527
Intra-regional	36 219 391	38 373 249	39 532 821	40 585 645	40 434 465
World total	135 359 229	143 616 274	149 168 510	150 855 623	149 215 384
	Percentage change				
Main East-West routes	4.03	5.6	4.9	-1.7	-0.3
Other routes (Non-mainlane)	1.40	6.5	3.2	3.0	-1.6
of which					
Non-mainlane East-West	2.57	5.8	0.0	4.8	-3.3
North-South	-0.37	5.6	2.9	0.1	-1.8
South-South	-1.68	8.9	7.4	3.9	-2.4
Intra-regional	2.75	5.9	3.0	2.7	-0.4

Source: UNCTAD secretariat calculations, based on data from MDS Transmodal, World Cargo Database, June 2021.
Note: Non-mainlane East West: Trade involving Western Asia and the Indian Sub-continent, Europe, North America, and East Asia.
North-South: Trade involving Oceania, Sub-Saharan Africa, Latin America, Europe, and North America.
South-South: Trade involving Oceania, Western Asia, East Asia, Sub-Saharan Africa and Latin America.
Intra-regional: Trade within Europe, Africa, Asia, North America, Latin America and Oceania.

Table 1.9 Containerized trade on major East-West trade routes, 2014–2021
(million TEU and percentage annual change)

	Eastbound	Westbound		Eastbound	Westbound		Eastbound	Westbound	
	East Asia–North America	North America–East Asia	Total Trans-Pacific	Northern Europe and Mediterranean to East Asia	East Asia to Northern Europe and Mediterranean	Total Asia-Europe	North America to Northern Europe and Mediterranean	Northern Europe and Mediterranean to North America	Transatlantic
2014	16.1	7.0	**23.2**	6.3	15.5	**21.8**	2.8	3.9	**6.7**
2015	17.4	6.9	**24.2**	6.4	15.0	**21.3**	2.7	4.1	**6.8**
2016	18.1	7.3	**25.4**	6.8	15.3	**22.1**	2.7	4.2	**6.9**
2017	19.3	7.3	**26.6**	7.1	16.4	**23.4**	2.9	4.6	**7.5**
2018	20.7	7.4	**28.0**	7.0	17.3	**24.3**	3.1	4.9	**8.0**
2019	19.9	6.8	**26.7**	7.2	17.5	**24.8**	2.9	4.9	**7.8**
2020	20.6	6.9	**27.5**	7.2	16.9	**24.1**	2.8	4.8	**7.6**
2021	24.1	7.1	**31.2**	7.8	18.5	**26.3**	2.8	5.2	**8.0**
	Percentage annual change								
2014–2015	7.5	-2.2	**4.6**	0.9	-3.2	**-2.0**	-3.1	5.1	**1.7**
2015–2016	4.3	6.6	**5.0**	6.3	2.4	**3.6**	0.2	3.2	**2.0**
2016–2017	6.6	-0.4	**4.6**	4.2	6.8	**6.0**	7.3	8.0	**7.7**
2017–2018	7.1	1.0	**5.4**	-0.9	5.7	**3.7**	5.3	7.6	**6.7**
2018–2019	-3.6	-7.4	**-4.6**	2.9	1.4	**1.8**	-4.7	-0.2	**-1.9**
2019–2020	3.2	1.6	**2.8**	-0.1	-3.7	**-2.6**	-4.6	-2.4	**-3.2**
2020–2021	17.1	2.7	**13.5**	8.0	9.5	**9.0**	1.4	9.0	**6.2**

Source: UNCTAD, based on MDS Transmodal, World Cargo Database, June 2021.

On other routes, the Asia-Europe trade declined by 2.6 per cent, reflecting reduced demand in Europe – despite frontloading and inventory building in the United Kingdom ahead of Brexit in 2020. And transatlantic trade fell by 3.2 per cent, depressed by reduced import demand from Europe, although solid import demand from North America moderated to 2.4 per cent the fall on the backhaul journey.

The crunch in container shipping in 2021 revealed many logistical problems, inefficiencies and vulnerabilities that are threatening the sustainability of the recovery and the competitiveness of supply chains. In May 2020, global schedule reliability had been 75 per cent, but in May 2021 it was only 39 per cent and in that month the average delay for late vessels was six days – down from the February peak of seven days, but still higher than that for most of 2020 (Metroshipping, 2021). At the same time, however, freight rates and surcharges, and fees, including demurrage and detention fees, had soared, though the latter rates were inconsistent across ports and carriers (Waters, 2021a).

These problems have been exacerbated by shipping network disruptions. In May 2021, the month-long closure of the port of Yantian in China increased cargo bottlenecks leading to a backlog affecting the region's manufacturing sector and increasing the number of blank sailings causing headaches for shippers (Port Technology International, 2021a; Waters, 2021b). Although less disruptive, the March 2021 grounding of the 20,150-TEU containership Ever Given in the Suez Canal blocked the canal, increasing delays for ships heading for Europe and added to a logistical disruption and port congestion. Some voyages had to be re-routed around the Cape of Good Hope, adding up to 7,000 miles to the journey – and pushing up freight and charter rates (Clarksons Research, 2021c).

Carriers argue that they are deploying all available capacity and that the current strain is being triggered by large and rapid swings in demand, and the surges in trade flows. This is leading to delays in returning containers and reducing effective capacity, making it difficult to cut delays, rates, and fees, while forcing carriers to adjust their networks and avoid some ports. They had already been advising customers on the Transpacific route, for example, that schedule disruptions would lead to blank sailings (Mongelluzzo, 2021a). As for terminal operators, they blame delays at ports on carriers, noting increases in double-sailings – two or more vessels sailing within the same week on the same service string or ordered set of ports. Large peaks and troughs in volumes leading to operational instability have disrupted operations and increased congestion (Waters, 2021c).

From their perspective, shippers have been looking for alternatives and solutions. Some have resorted to higher-priced air freight, while on the Far East-Europe route they have also been attracted by rail transport. According to Chinese customs data, rail volumes and capacity are still relatively small, but the two-way trade value nearly trebled in the first five months of 2021 (Global Times, 2021). Meanwhile, on the Transpacific and intra-Asian routes some commodities, such as grain and forestry goods, have seen a temporary de-containerization with goods despatched on dry bulk ships, adding to the demand for multipurpose ships and dry bulk carriers (Sand, 2021c).

To secure space on vessels, some shippers are seeking longer-term, multi-year, end-to-end contracts with carriers. For their part some carriers seek to convert 'ocean customers' to long-term 'end-to-end logistics customers'. Under these arrangements, shippers have access to logistics services such as warehousing, customs clearance, visibility, and the ability to speed up or slow down shipments (Knowler, 2021). Examples include Maersk's aim to become a full-service, end-to-end integrator, and the focus of CMA CGM and its CEVA Logistics division on creating integrated services (Tirschwell, 2021). In response to increasing congestion and shrinking ocean capacity Maersk has launched the first block train intermodal service between Europe and China (Port Technology International, 2021b).

The Global Shippers' Forum argues that the real crunch point for shippers is the plummeting service performance and the unpredictability of container delivery, and has renewed its call to remove the consortia block exemption regulation (Baker, 2021a). It points to the increasing number of blank sailings – ships skipping a port or ports, or cancelling the entire string – which reduce the number of containers that shippers can export. This disproportionately affects lower-paying shippers since carriers favour cargo from higher-paying customers (Waters, 2021c). In this respect, the United States Congress is drafting legislation to strengthen the Federal Maritime Commission's oversight of carriers' shipping practices (Gallagher, 2021).

The Global Shippers Alliance maintains that since no carrier on its own will be able to guarantee good connectivity and port pairs, the current supply chain crisis is unlikely to be solved by further regulation of container shipping. Instead it calls on carriers to take more risk, building contingency into their prices and employing new technology to make supply chain forecasts more accurate and more transparent. The best solution, they say, would be to adopt enforceable contracts, which would also act as hedges against uncertainty and enhance collaboration among shippers, carriers and forwarders (Baker, 2021b).

Overall, since early 2020, when the pandemic first hit, the narrative for container shipping has thus shifted dramatically. Carriers have been able to manage ship capacity so as to mitigate initial disruptions but port and landside businesses required more time to adjust their yard and gate operations which often led to inefficiencies in terminal operations, such as the management of container stacking (Notteboom, Pallis and Rodrigue, 2021).

Shippers are caught in this storm and need to better manage their supply chains and adapt to lower capacity (Drewry Maritime Research, 2021d). They should adopt proactive supply chain strategies that anticipate delays and promote visibility. While some carriers and ports (e.g., Maersk and DP World) are emerging as end-to-end integrators, they should spare no effort to address congestion and service reliability and ensure that maritime trade is not undermined by the current logistical hurdles.

Meanwhile in mid-2021 pressure in container shipping continued unabated, with shippers increasingly worrying about the reliability of services and their ability to secure space for their shipments. On 9 July 2021, the President of the United States signed an executive order that encourages the United States Federal Maritime Commission "to ensure vigorous enforcement against shippers charging American exporters exorbitant charges" (Holt, 2021). Since then, Federal maritime regulators have ordered eight container lines to provide details showing how congestion port surcharges meet legal and regulatory requirements (Szakonyi, 2021).

5. Container port traffic disrupted as congestion heightens and shipping adjusts operations and schedules

For ports, the years 2020 and 2021 were highly disruptive. In 2020, global container port throughput fell by 1.2 per cent, to 815.6 million TEU (table 1.10). For 2021, however, volume is projected to grow by 10.1 per cent as the global economy and trade recover, along with increasing optimism arising from the vaccine rollout (Drewry Maritime Research, 2021e). But some ports fared better than others. Antwerp, for example, fared much better in the COVID-19 crisis than it had during the 2009 downturn.

In 2020, Asia, with nearly two-thirds of the throughput, maintained its position as the global hub for container port traffic (figure 1.10). Europe was the second-largest container port handling region in 2020 (14.4 per cent). Together, North America (7.5 per cent), Latin America and the Caribbean (7.2 per cent), Africa (4.0 per cent), and Oceania (1.6 per cent) accounted for the remaining shares. North America and Asia benefited from the swift trade rebound in the second half of 2020, but recurrent virus outbreaks and pandemic containment measures, among other factors were a drag on container port traffic in Europe and other regions.

China's dominance is also evident from data on the world's top 20 ports around half of which are in China (figure 1.11). In 2020, cargo throughput in these leading ports declined, though there were some exceptions, notably Tanjung Pelepas with growth of 7.7 per cent and Long Beach which benefited from a surge in the United States containerized imports. In the fourth quarter of 2020, volumes at Long Beach rose 23 per cent. Los Angeles also enjoyed 22 per cent growth in the last quarter of the year but still closed the year down 1.3. per cent.

Table 1.10 World container port throughput by region, 2019–2020
(million TEU and annual percentage change)

	2019	2020	2019–2020
Asia	534.8	532.7	-0.4%
Africa	32.5	32.5	0.0%
Latin America and the Caribbean	60.1	59.0	-1.8%
Europe	122.6	117.4	-4.2%
North America	62.4	61.2	-1.9%
Oceania	12.9	12.8	-0.8%
World Total	**825.3**	**815.6**	**-1.2%**

Source: UNCTAD secretariat based on data collected by various sources, including Lloyd's List Intelligence, MDS Transmodal, Dynamar B. V., Drewry Maritime Research, Professor Jean-Paul Rodrigue, Hofstra University, as well as information published on relevant port authorities and container port terminals websites. In some cases, data was estimated based on liner shipping connectivity data at country level.

Note: Data reported in the format available. In some cases, country volumes were estimated based on secondary source information and reported growth rates. Country totals may conceal the fact that minor ports may not be included. Therefore, in some cases, data in the table may differ from actual figures.

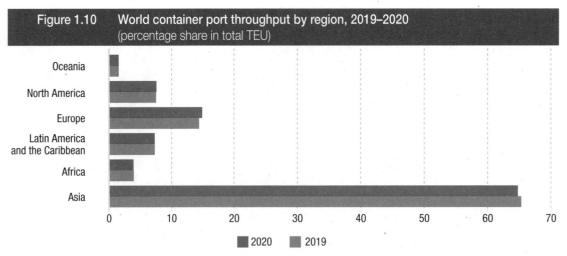

Figure 1.10 World container port throughput by region, 2019–2020
(percentage share in total TEU)

Source: UNCTAD secretariat calculations, derived from table 1.10.

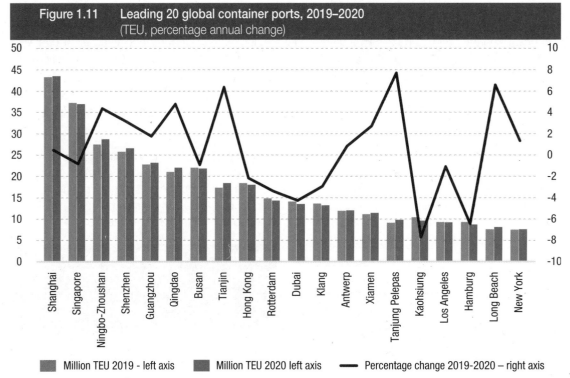

Figure 1.11 Leading 20 global container ports, 2019–2020
(TEU, percentage annual change)

Source: UNCTAD based on data published on Hamburg Port Authority website (www.hafen-hamburg.de/en/statistics/top-20-container-ports), accessed July 2021.

Nearly all leading Chinese ports increased their throughput. Shanghai saw slow growth but remained the world's leading port, while growth in Tianjin was 6.4 per cent and Qingdao 4.8 per cent. In Europe and North America port performance varied. Outside this group, the fall in throughput in Colombo was caused by pandemic-induced labour shortages and limited capacity on mainline vessels. Beirut continued to lose traffic to Tripoli following the 2020 port explosion (Drewry Maritime Research, 2021f).

New York (+1.3 per cent) and Antwerp (+0.8 per cent) have been more resilient, while Kaohsiung (-7.7 per cent) and Hamburg (-6.5 per cent) were severely hit. Others such the ports of Dubai (-4.3 per cent), Rotterdam (-3.4 per cent), Klang (-2.9 per cent), and Busan (-0.9 per cent), recorded drops in volumes handled.

The COVID-19 pandemic was a big disruptor that has created challenges but also opportunities for the sector. Digitalization and environmental sustainability have become key pillars of the post-pandemic recovery. Industry and governments are considering opportunities that may arise from 'building back better'. For example, in 2021 COSCO Shipping Ports launched a green finance framework to drive green and smart port development (Greenport, 2021a). Elsewhere, the European Union granted €25 million to a consortium led by the Port of Rotterdam to run pilot projects on sustainable and smart logistics. Project partners will also design and implement digitalization and automation solutions for the energy transition (Greenport, 2021b). Meanwhile, the United States' $1.9-trillion spending plan includes funds earmarked for transport infrastructure and resilience, including ports (Port Strategy, 2021).

B. OUTLOOK AND LONGER-TERM TRENDS

As the global economy moves towards its next normal, there are optimistic signs for maritime trade. Some of the pandemic's impacts and legacies could linger, but the short-term outlook is generally positive.

1. A positive short-term outlook but with risks and uncertainties

Global economic prospects improved by late 2020, supported by vaccine rollout in advanced regions, the possibility of additional spending in some major economies, and the easing of containment measures and restrictions in some parts of the world. While emerging trends are encouraging, uncertainty remains as the sustainability of the nascent, fragile and divergent recovery depends on the pandemic's path and a broader rollout of vaccines worldwide.

UNCTAD projects shipping volumes to increase by 4.3 per cent in 2021, and exceed their 2019 levels (table 1.11). Containerized trade is expected to grow by 7.7 per cent. Over the 2022–2026 period, total maritime trade is expected to grow 2.4 per cent annually – compared with 2.9 per cent over the previous two decades. Maritime trade is projected to moderate along with GDP (IMF, 2021).

The intensified cost pressures, inefficiencies, and vulnerabilities in the maritime supply chain, driven primarily by the COVID-19 disruption and its knock-on effects on shipping and ports, could continue to disrupt supply chains, raising both production costs and consumption prices. But these pressures are expected to ease when global demand patterns are normalized, manufacturing capacity comes online, and logistical assets are optimized to improve the balance between supply and demand.

A further concern is trade protectionism and trade tensions between China and its trading partners, including the United States and Australia. Governments may also resort to trade protectionism to mitigate discontent and social tensions arising from the impact of COVID-19 on employment and social inequalities.

On the upside, the recovery should be driven by fiscal support measures, though there are uncertainty regarding the duration of current stimulus packages and government spending while developing countries continue to be under pressure – having limited fiscal policy space and low access to vaccines.

Other positive trends include the signing in 2020 of the Regional Comprehensive Economic Partnership and the coming into force of the African Continental Free Trade Area (AfCFTA) in 2021. UNCTAD expects the AfCFTA to boost intra-African trade by about 33 per cent and cut Africa's trade deficit by 51 per cent (Saygili, Peters, Knebel, 2018). AfCFTA also has important implications for maritime transport and services trade (box 1).

Table 1.11 International maritime trade developments forecasts, 2021–2026
(annual percentage change)

	Annual Growth	Years	Seaborne trade flows
UNCTAD	4.3	2021	Total seaborne trade volume
	3.2	2022	
	2.4	2023	
	2.3	2024	
	2.3	2025	
	2.2	2026	
UNCTAD	7.7	2021	Containerized trade volume
	5.9	2022	
	4.7	2023	
	4.4	2024	
	4.2	2025	
	4.1	2026	
Clarksons Research, *Seaborne Trade Monitor*, June 2021	4.3	2021	Total seaborne trade volume
	3.1	2022	
	5.9	2021	Containerized trade volume
	4.0	2022	

Source: UNCTAD secretariat based on own calculations and forecasts published by the indicated institutions and data providers.

Note: Projections are based on the estimated elasticities of maritime trade with respect to world GDP, export volumes, investment share in GDP for the 1990-2020 period as well as monthly seaborne trade data published by Clarksons Research.

> **Box 1**   Implications of AfCFTA for maritime transport in Africa
>
> The African Continental Free Trade Area (AfCFTA) agreement entered into force in 2019, and its implementation commenced in 2021. It aims to increase intra-African trade by eliminating import duties, and to double this trade if non-tariff barriers are also reduced. Adequate transport infrastructure and services in Africa, including maritime transport connectivity, are critical to the full realization of the benefits of AfCFTA. Moreover, the AfCFTA is expected to increase demand for different modes of transport, including maritime transport, which in turn will increase investment requirements for infrastructure and equipment – ports and vessels in the case of maritime transport.
>
> The Services Protocol of AfCFTA sets out principles for enhanced continental market access and services- sector liberalization. The five priority sectors identified include transport, business services, communication services, financial services, and tourism. AfCFTA could therefore be a game-changer for investment in transport infrastructure and services. Maritime transport infrastructure in Africa includes several ports across the continent which landlocked countries access through road and rail corridors. Some of these ports are congested and located in the middle of cities.
>
> A study by the Economic Commission for Africa, with a time horizon of 2030, provides a forecast of the requirements for transport infrastructure, services, and equipment as a result of the implementation of AfCFTA. The analysis shows that in 2019, maritime transport accounted for almost a quarter of total intra-African freight transport demand (22 per cent). It indicates that the number of tons transported by vessels with the implementation of AfCFTA would increase from 58 million to 132 million tons. The total maritime transport share is expected to increase only by 0.6 per cent, from 22.1 per cent to 22.7 per cent in the scenario where AfCFTA and priority infrastructure projects are implemented, and by 1.5 per cent in the scenario where AfCFTA is implemented but priority infrastructure projects are not implemented. If priority infrastructure projects are implemented some traffic is expected to shift to rail and road as these projects focus mainly on road and rail transport.
>
> The study shows that countries in different subregions of the continent will experience a surge in traffic through their ports by 2030 owing to AfCFTA, including Gabon (Central Africa), Ghana, Gambia (West Africa), Somalia, Comoros, Mauritius (East Africa) and Mozambique, Madagascar, Namibia (Southern Africa). The study estimates the required size of Africa's maritime transport fleet due to the implementation of AfCFTA. In this regard, in the scenario where AfCFTA is not implemented and no priority infrastructure projects are implemented by 2030 compared to 2019 (the baseline), the size of the fleet is estimated to increase by 43 per cent for bulk and 40 per cent for container cargo. However, compared to 2019 and to satisfy intra-African trade demand, the size of the fleet for bulk and container cargo is estimated to increase by 200 per cent if AfCFTA is implemented and no infrastructure projects are executed. In the scenario where AfCFTA and the different infrastructure projects are implemented by 2030, the fleet is estimated to increase by 188 per cent for bulk and 180 per cent for container cargo.
>
> The most significant vessel demand to support trade flows resulting from AfCFTA, compared to the baseline of 2019, is within North Africa (35 per cent of the total vessel fleet), from North Africa to East Africa (15 per cent), and from North Africa to West Africa (11 per cent). It is worth noting that the second priority action plan of the Programme for Infrastructure Development in Africa (PIDA PAP II), endorsed by the Summit of African Union Heads of State in February 2021, and to be implemented between 2021 and 2030, recognizes the importance of maritime transport to Africa's socio-economic development and regional integration. In this regard, PIDA PAP II includes the following projects:
>
> - Maritime connectivity between the islands of Comoros;
> - Construction of petroleum jetty and associated storage facilities at Albion, Mauritius; and
> - Praia-Dakar Shipping and Maritime Services Project.
>
> *Source:* Economic Commission for Africa (forthcoming). *Implications of the African Continental Free Trade Area for Demand of Transport Infrastructure and Services.* Addis Ababa, Ethiopia.

2. Long-term outlook shaped by structural factors and lingering effects of the pandemic

The long-term outlook will be shaped by a range of continuing structural trends. These include changing patterns of globalization, the drive for more-resilient supply chains, changes in consumer spending and the growth of ecommerce, the need for environmental sustainability, the global energy transition, and the continuing uptake of digitalization.

Shift in globalization patterns

Even before the COVID-19 pandemic, global value chains were being increasingly shaped by rising demand and new industry capabilities in the developing regions, and growth in automation and robotics, the shift from tradeable goods to service, and limited growth in vertical specialization and global fragmentation of production that reflect maturing value chains in China and the United States. The hyper-globalization of the late-1990s and early-2000s appears to be decelerating. Enterprises, particularly in automotive, computer and electronics industries, are aiming to locate production closer to demand and consumption markets. Developing countries are increasingly consuming their own products and reducing their imports of intermediate goods while creating more comprehensive domestic supply chains (UNCTAD, 2019).

Decisions will also be shaped by recent episodes of shipping network disruption (Suez Canal blockage, surge in COVID-19 cases in South China), chip shortages that close car manufacturing, shipping delays and soaring costs. Existing shifts in globalization patterns can be expected to accelerate (Yap and Huan, 2018).

Some countries are also aiming for greater self-reliance particularly in goods considered to be strategically valuable, such as pharmaceuticals and medical equipment, and new technology (Fitch Solutions, 2020). This is illustrated by initiatives such as Made in China 2025, Buy American, Strategic Autonomy in Europe, and Self-sufficient India – as well as incentives to move supply chains closer to home in Japan, the Republic of Korea and Taiwan Province of China.

In the United States, the new administration has already indicated its intention to build supply chains that rely less on China for strategically important products (Wood and Helfgott, 2021). And in China the recent 14th Five-Year Plan is expected to boost domestic consumption and expand the domestic market for China's manufactured goods. It also seeks to achieve technological self-sufficiency and expand exports (Fitch Solutions, 2021). Overall, the plan is expected to benefit shipping while promoting energy, grains, minor bulk commodities, and chemicals imports.

While the pandemic could deepen pre-existing changes to globalization patterns, it has also reaffirmed China's important role in sustaining international trade. With around one-third of global trade, China is showing the resilience and determination to remain the 'factory of the world'. West and South Asia, South America, Western Europe and the Mediterranean regions recorded export growth in the fourth quarter of 2020, although of a lower scale (Teodoro, 2021).

Since 2018 the United States has increased tariffs, but rather than inducing a return of production to the United States this tended to shift manufacturing within Asia. In 2020 Cambodia, for example, took over a large part of China's market share in United States imports of Christmas lights. During the same period, exports of bikes to the United States from Cambodia jumped by 478 per cent and from Taiwan Province of China by 30 per cent. Tariffs have not provoked a large-scale nearshoring and have had little impact on ton-miles as containerized exports from China or neighbouring East Asian countries hardly affect the distances travelled to the United States (Sand, 2020b).

Nevertheless, while China continues to lead world exports its predominance can be expected to moderate as its economy matures and relies more on domestic than external demand. This implies that imports in value terms are likely to increase faster than exports (Nicita and Razo, 2021), suggesting potential shifts in shipping patterns and trade, and changes in maritime transport demand.

Nevertheless an outright reversal of globalization will be difficult. Global supply chains are the product of years of investment, relationship-building, and knowledge acquisition, and China's large production and logistical capacity and economies of scale are difficult to replace. This was demonstrated by the increased imports of electronics in 2020, which triggered a shift of some production and sourcing back to China. And while imports of machinery and electrical equipment, and computers from Mexico may have increased over recent years, often components are exported from China to Mexico for assembly in manufacturing facilities near the United States border (Cassidy, 2021a).

It may be fairly straightforward to change labour-intensive and low-value supply chains. Apparel and textiles, for example, are already moving away from China to Bangladesh, Viet Nam, and Ethiopia. Turkey is also a major producer of clothing, shipping goods to Europe. But it is more complex for mid- and high-value-added manufacturing. For semiconductors, for example, one study estimated that only 9 to 19 per cent of trade flows could potentially shift. For car exports the estimate was 15 to 20 per cent though for pharmaceuticals it was 38 to 60 per cent (Lund et al., 2020).

Some companies are nevertheless aiming to diversify production sites, with a 'China +1' strategy and will continue to look for alternative sources which will require adjusting networks and inventory management strategies and transport and shipping routes. This is resulting in new trade flows as observed in the case of China-Mexico-United States, or from other countries in East Asia to the United States. Morocco, and Central and Eastern Europe can be expected to strengthen their position as new suppliers to the North American and European consumer market, for cars, electronics, and heavy machinery (Fitch Solutions, 2020). In the long term, automation could make reshoring and nearshoring more economically viable.

The pandemic and its fallout are likely to hasten this transition, but the outcome will likely be a blended approach, balancing localized and global sourcing depending on product and geography (UNCTAD, 2021c). These trends have major implications for maritime transport, as carriers need to redefine distances and routes and offer more flexible shipping services. A reconfiguration of supply chains has implications for vessels, sizes, ports of call, and distance travelled.

Mainstreaming supply chain resilience, risk assessment and preparedness

Over the years, global supply chains have become more sophisticated and extensively interlinked. They have also become vulnerable to wide-ranging risks, with more potential points of failure. This became clear from the COVID-19 disruption which tested existing supply chains and logistics networks and their underlying business models.

Aiming for greater supply chain resilience will mean diversifying business partners and suppliers, improving forecasting of demand and volumes, ensuring better management of inventories and safety stocks, and carefully rethinking the trade-offs between just-in-time and just-in-case supply chain business models (Cassidy, 2021b). While responses may be influenced by sentiment at the height of the pandemic, over 90 per cent of the supply chain executives that had responded to a May 2020 survey, were planning to enhance resilience (Lund et al., 2020). This can be achieved, for example by allowing for redundancy across suppliers, nearshoring, regionalizing their supply chains, dual-sourcing raw materials, backing up production sites, increasing inventory of critical products, strengthening supply-chain risk management, improving end-to-end transparency, and minimizing exposure to cybersecurity and other shocks.

Investors, rating agencies, and regulators increasingly expect ports and shipping companies to integrate risks into their plans (Kim and Ross, 2019). For this they will need to devise and implement risk management and business continuity strategies, and ensure visibility across extended supply networks, while building strong relationships with key partners, including shippers and inland transport providers. To this end, they can use new technologies that enable end-to-end visibility, collaboration, responsiveness, agility, and optimization of operations (Koch, Vickers, and Ritzmann, 2020). It will also be important to support the digitalization of smaller ports and inland terminals (Schwerdtfeger, 2021a).

Any effort to strengthen the resilience of the maritime supply chain would be in vain if the human resources and labour dimension is not addressed as a matter of priority. The pandemic has underscored the critical role of seafarers. Smooth delivery of trade by shipping and efficient handling of cargo by ports depend mainly on their labour forces. Crew members need to rotate at the end of their contract periods. At the height of the disruption, hundreds of thousands of seafarers could not be repatriated, while an equivalent number were stuck at home and could not join their ships and provide for their families. As indicated in Chapter 5, the shipping industry has asked that vaccines be secured and allocated specifically for seafarers. In May 2021, the International Maritime Organization called on Member States to support the fair global distribution of COVID-19 vaccines.

In support of these efforts, Singapore, as a global hub port and international maritime centre, is considering providing vaccines to crews on vessels calling at its port (Ang, 2021). Elsewhere, in June 2021 the Royal Association of Netherlands Shipowners launched the Vaccination Programme for Seafarers.

The growth in ecommerce and change in consumption patterns

Pandemic-induced shifts in consumption and shopping habits together with digitalization have accelerated growth in ecommerce. In 2019, around 16 per cent of retail sales were online, a proportion which grew in 2020 to 19 per cent (UNCTAD, 2021b). UNCTAD estimates the global ecommerce market in 2019 at $27 trillion, equivalent to 30 per cent of GDP. Ecommerce fulfilment provides new business opportunities – in particular for warehousing and distribution facilities at seaports, inland rail hubs, and near airports. This can reduce supply chain uncertainties enabling retailers to keep more inventory at hand. Retailers are also seeking properties with large container yards to store containers on chassis (Mongelluzzo, 2021b).

Ports close to, or well-connected to, large population centres could tap this business potential (Drewry Maritime Research, 2021d). Already, some container shipping companies and ports are positioning themselves to emerge as door-to-door service integrators (e.g., Maersk and DP World). Container shipping companies have recently invested in other parts of the supply chain, including warehousing, aircraft, and distribution (Steer and Dempsey, 2021).

The imperative for environmental sustainability and the energy transition

The COVID-19 pandemic has increased the focus on environmental sustainability. Maritime transport is facing growing pressure to decarbonize and enable an effective energy transition – both as a transporter and user of energy. Fossil fuels make up over one-third of global maritime trade but demand for these fuels is expected to fall, with clear implications for tankers and coal carriers, while demand is likely to increase for ships transporting hydrogen or ammonia.

At the same time ships are also expected to shift their own fuel mix and use new ship designs to cut fossil fuel consumption and reduce carbon emissions. To mitigate these additional costs, shipping is set to rely on technological and operational adjustments.

Ports are also expected to play their part and become smart and green. Some governments have earmarked some of the pandemic-induced stimulus packages for smart and green maritime transport projects.

Acceleration in digitalization

Port authorities, shippers, and freight forwarders that had invested in digital infrastructure and connectivity and promoted data exchange navigated more smoothly through the COVID-19 disruption (Schewerdtfeger, 2021a). But this also widened the digital divide between developed and developing regions. Countries that were less advanced were less able to mitigate the pandemic and diversify their economies.

UNCTAD expects the fast shift towards digitalization to strengthen the market positions of a few digital mega platforms. If left unaddressed, the yawning gap between under-connected and hyper-digitalized countries will widen, exacerbating inequalities (UNCTAD, 2020b).

Investing in digital infrastructure is crucial for information sharing and effective resource planning. Automation and smart technologies, including artificial intelligence, can solve many of the challenges faced by the industry, such as how to process more cargo in an environmentally friendly manner (Schewerdtfeger, 2021c). Developing countries should be supported in their efforts to implement digital tools to advance environmental sustainability, economic efficiency, and resilience.

C. POLICY CONSIDERATIONS AND ACTION AREAS

Against the backdrop of an already more challenging global geopolitical and trade policy landscape, the COVID-19 disruption shone light on the vulnerabilities of the global supply chains, including their underlying maritime transportation networks. Governments are forging ahead with 'build back better' policies and initiatives to ensure that risks, environmental sustainability, and technology are integrated as pre-requisites for a sustainable and resilient post-pandemic world. While maritime trade is currently in recovery mode, the pandemic is having a lasting impact. The recovery is uneven and fragile and some pre-existing trends are being amplified or accelerated.

Maritime transport and trade are at the forefront of these trends, and the following priority actions areas will help the sector navigate through the transition:

- *Vaccination* – Strengthen international efforts to tackle the pandemic and ensure wider vaccination across regions and within the shipping industry, with vaccination plans for seafarers topping the priority list. A two-paced vaccine approach widening the gap between countries, populations and economic sectors will perpetuate asynchronous recovery patterns, which may have proved helpful in preventing a protracted downturn when the pandemic hit but raises concerns about the sustainability of the recovery. A multi-paced vaccine-led recovery entails risks, and would exacerbate inequalities which could culminate in social tensions and disruptions. The International Monetary Fund estimates that $50 billion is required to end the pandemic across the world and ensure that vaccines are accessible to developing countries. The dividend for the world economy extends beyond saving lives, as investing in global vaccination plans could accelerate economic recovery and generate some $9 trillion in additional global output by 2025 (Georgieva et al., 2021).

- *Digital divide* – Help countries and their maritime industries to catch up and close the digital gap. The pandemic may have exacerbated the digital divide between developed and developing regions and between the hyperconnected and weakly connected. Closing the gap is important and could form part of relevant post-pandemic recovery plans and other support measures.

- *Facilitate trade* – The wheels of trade and shipping kept the world going when the pandemic hit and helped lift the world economy. Going forward, trade should be further enabled by adopting supportive policy measures that minimize trade restrictiveness and protectionist tendencies.

- *Fiscal support* – Carefully time the winding up and withdrawal of fiscal support measures, to avoid a premature withdrawal that stifles the nascent recovery. For most developing countries where fiscal measures similar to those in developed regions could not be deployed, international cooperation and targeted aids are becoming crucial.

- *Stakeholder collaboration* – Stakeholders in the maritime supply chain, including carriers, ports, inland transport providers and shippers, should work together to ensure that maritime transport remains a reliable, predictable, and efficient mode of transport that links supply chains and enables trade. And to ensure visibility and transparency they should ensure enhanced communications, and sharing of data and information.

- *Ecommerce* – Shipping and ports should explore the business opportunities arising from growth in ecommerce, accelerated digitalization and the growing environmental sustainability imperative, and take these opportunities to promote profitability while also providing quality services that meet customer and supply chain requirements.

- *Sustainability* – Expand efforts to promote environmental sustainability as part of the various stimulus packages and post-pandemic recovery plans. Support for decarbonization under the IMO framework should not waver, while ensuring that the implications for developing countries are well understood.

- *Energy transition* – Promote investment in fleets, technologies, and infrastructure, including ports and hinterland connections, to support a maritime supply chain energy transition and environmental sustainability.

- *Resilience building and future proofing* – Prioritize preparedness, risk management, digitalization, environmental sustainability, and improving data and forecasting. End-to-end visibility will increase resilience while enhancing efficiency and productivity gains. A portfolio of measures can improve resilience including redundancy across suppliers, dual-sourcing, backing up production sites, and managing inventory, and stocks, along with risk management, and end-to-end transparency. Hybrid solutions can also be envisaged, involving extended supply chains with an element of nearshoring and reshoring.

REFERENCES

Ang I (2021). Singapore looks to vaccinate foreign seafarers during crew changes. 8 July. Tradewinds.

Baker J (2021a). Shippers renew call for block exemption to be removed. Lloyd's Loading List. 14 June. www.lloydsloadinglist.com.

Baker J (2021b). Shippers seek collaboration, not regulation, to fix supply chain. Lloyd's Loading List. 23 July. www.lloydsloadinglist.com.

Clarksons Research (2020). LNG Trade and Transport 2020. London.

Clarksons Research (2021a). Shipping Review Outlook. Spring.

Clarksons Research (2021b). *Seaborne Trade Monitor*. Volume 8, No.6. June.

Clarksons Research (2021c). Suez Canal Blockage: Summary and Context. Briefings. 29 March.

Connolly M (2021). West coast Latin America looks to Asia for refined products. Argus Media. 28 June.

Danish Shipping Finance (2021). Shipping Market Review. May.

Danish Shipping Finance (2020). Shipping Market Review. November.

Drewry Maritime Research (2021a). Shipping Insight. April.

Drewry Maritime Research (2021b). Shipping Insights. February.

Drewry Maritime Research (2021c). Shipping Insights. May.

Drewry Maritime Research (2021d). Ports and Terminals Quarterly Insights. Quarter 2. May.

Drewry Maritime Research (2021e). Container Forecaster. Second Quarter. June.

Drewry Maritime Research (2021f). Port Quarterly Insight. Quarter 1. March.

Ebregt J (2020). The CPB World Trade Monitor: Technical description (update 2020). CPB Background Document. CPB Netherlands Bureau for Economic Policy Analysis. The Netherlands

Economic Commission for Africa (forthcoming). *Implications of the African Continental Free Trade Area for Demand of Transport Infrastructure and Services*. Addis Ababa, Ethiopia.

Economist Intelligence Unit (2021). Trade in Transition. Global Report. DP World.

Fitch Solutions (2020). Deglobalization Will Prompt Lasting Changes to Supply Chains.9 October.

Fitch Solutions (2021). China's 14th Five-Year Plan: Evolution of Policy to Meet Challenges. March 2021.

Gallagher J (2021). Congress drafting law barring ocean carriers from refusing US exports. Freight Waves. 15 June. https://www.freightwaves.com.

Georgieva K, Ghebreyesus TA Malpass D and Okonjo-Iweala N (2021). A New Commitment for Vaccine Equity and Defeating the Pandemic. 1 June 2021. www.imf.org.

Greenport (2021a). Green finance initiative launches. 13 May 2021.

Greenport (2021b). Greenport project funding. 12 May 2021.

Global Times (2021). China-Europe freight trains jump 96 per cent in Jan-Feb, a steady pillar for BRI trade in hard times. 8 March.

Hellenic Shipping News (2020). Guinea set to supply iron ore from 2026. 1 June.

Holt G (2021). Biden executive order could bring more scrutiny to rising container shipping costs. SP Global. 9 July. www.spglobal.com.

IMF (2021). World Economic Outlook: Managing Divergent Recoveries. April. Washington. www.imf.org.

Jack E and Don C (2021). Lack of Tiny Parts Disrupts Auto Factories Worldwide. 13 January. *New York Times*.

Knowler G (2021). Maersk signals next logistics acquisition will be billion-dollar deal. 11 May. www.JOC.com.

Koch M, Vickers P and Ritzmann S (2020). Building Supply Chain Resilience beyond COVID-19. Deloitte.

Lund S et al. (2020). Risk, resilience, and rebalancing in global value chains. McKinsey Global Institute. August. www.mckinsey.com.

MDS Transmodal (2021). World Cargo Database. 16 June. www.mdst.co.uk.

Metroshipping (2021). Container vessel schedule reliability at all-time lows on a global level. 8 July. www.metroshipping.co.uk.

Mongelluzzo B (2021a). US port delays force 'structural' blank sailings on Asian services. 26 January. Journal of Commerce. www.JOC.com.

Mongelluzzo B (2021b). Record US warehousing demand to continue through 2021. 10 March. Journal of Commerce. www.JOC.com.

Nicita A and Razo C (2021). China: The rise of a trade titan. 27 April.

Notteboom T (2021). PortEconomics.eu. 2 March.

Notteboom T, Pallis A and J-P Rodrigue (2021) "Disruptions and Resilience in Global Container Shipping and Ports: The COVID-19 Pandemic vs the 2008-2009 Financial Crisis", Maritime Economics and Logistics, vol. 23, issue 2, No 1, 179-210, available at https://link.springer.com/article/10.1057%2Fs41278-020-00180-5.

PortStrategy (2021). Thinking Big. 24 May. www.portstrategy.com.

Port Technology International (2021a). Blank sailings at Yantian rise by 300 per cent as expert warns of "massive headaches" for global economy. 20 June. www.porttechnology.org.

Port Technology International (2021b). Maersk responds to ocean congestion with new intermodal route. 21 June. www.porttechnology.org.

Sand P (2020a). Phase One' agreement boosts exports and doubles volumes from last year. 15 December. BIMCO.

Sand P (2020b). US-China trade war reinforces change in Asia's manufacturing landscape. BIMCO. 23 September.

Sand P (2021a). Macroeconomics: shipping enjoys a higher trade multiplier as 2021 promises a slow recovery. BIMCO. 24 February.

Sand P (2021b). Imbalance on transpacific trade increases even as surge in US imports eases. BIMCO. 24 March.

Sand P (2021c). Container shipping: continued disruption will ensure carrier profitability well into 2021. BIMCO. 25 February.

Saygili M, Peters R, Knebel C (2018). African Continental Free Trade Area: Challenges and Opportunities of Tariff Reductions. UNCTAD Research Paper No. 15. UNCTAD/SER.RP/2017/15/Rev.1. February.

Steer G and Dempsey H (2021). Shipping groups prepare for more 'black swan' events after Suez blockage. 7 April. www.ft.com.

Schwerdtfeger M (2021a). CTAC 2021: Digitalise inland terminals to boost supply chain resiliency. 19 May.

Schewerdtfeger M(2021b). Smart technologies remain critical to solving ports' major challenges. 13 May.

Schewerdtfeger M (2021c) US must invest in digital infrastructure to avoid future supply chain cries. 17 May. Journal of Commerce. www.JOC.com.

Szakonyi M (2021). FMC orders carriers to justify US port congestion surcharges. 4 August. Journal of Commerce. www.JOC.com.

Teodoro A (2021). Changing lanes: China retains 'factory of the world' crown amid pandemic. 3 March. Available at https://lloydslist.maritimeintelligence.informa.com.

Tirschwell P (2021). Maersk upping ante on integrator strategy. 14 May. www.JOC.com.

UNCTAD (2019). *Review of Maritime Transport 2018*. Geneva and New York. Available at https://unctad.org/system/files/official-document/rmt2018_en.pdf.

UNCTAD (2020a). *Review of Maritime Transport 2020*. Geneva and New York. Available at https://unctad.org/webflyer/review-maritime-transport-2020.

UNCTAD (2020b). The COVID-19 Crisis: Accentuating the Need to Bridge Digital Divides. Digital Economy Update. UNCTAD/DTL/INF/2020/1. April. Available at https://unctad.org/system/files/official-document/dtlinf2020d1_en.pdf.

UNCTAD (2021a). Global Trade Update. World trade rebounds to record high in Q1 2021. May. Available at https://unctad.org/system/files/official-document/ditcinf2021d2_en.pdf.

UNCTAD (2021b). Estimates of Global E-Commerce 2019 and Preliminary Assessment of COVID-19 Impact on Online Retail 2020. Geneva. 3 May. Available at https://unctad.org/system/files/official-document/tn_unctad_ict4d18_en.pdf.

UNCTAD (2021c). Impact of the COVID-19 Pandemic on Trade and Development: Transitioning to a New Normal. UNCTAD/OSG/2020/1. Geneva. Available at https://unctad.org/system/files/official-document/osg2020d1_en.pdf.

UNDESA (2021). World Economic Situation and Prospects as of Mid-2021. United Nations. New York. Available at https://www.un.org/development/desa/dpad/wp-content/uploads/sites/45/publication/WESP2021_UPDATE.pdf.

Vanham P (2019). A brief history of globalization. World Economic Forum. 17 January. www.weforum.org.

Waters W (2021a). Demurrage and detention charges double in a year. Lloyd's Loading List. 29 June.

Waters W (2021b). South China port disruptions 'could extend to Christmas. Lloyd's Loading List. 7 July.

Waters W (2021c). Sharp increase in 'double-sailings' adds to pressure on ports. Lloyd's Loading List. 22 June.

Waters W (2021d). Blank sailings point to ocean space shortages well into the summer. Lloyd's Loading List. 24 May.

William C (2021a). Top US Shippers: Electronics importers return to China during COVID-19. 3 June. Journal of Commerce.

William C (2021b). Outlook 2021: Shippers face new twists, risky turns in pandemic-altered market. 4 January. Journal of Commerce.

World Trade Organization (2021). World trade primed for strong but uneven recovery after COVID-19 pandemic shock. Press/876. 31 March.

World Trade Organization (2020). Trade set to plunge as COVID-19 pandemic upends global economy. PRESS 855. 8 April.

Wood D and Helfgott A (2021). Inclusive and transparent dialogue can fix fragile US supply chains. 23 May. The World Trade Organization (2021). World trade primed for strong but uneven recovery after COVID-19 pandemic shock. Press/876. 31 March.

Yap J and Huan A (2018). The Good, the Bad, and the Ugly of Globalization. SIDS Directors Bulletin. Quarter 3. www.deloitte.com.

Yoon K and Lindsay R (2019). Ports: An industry guide to enhancing resilience. Resilience Primer. Four Twenty-Seven Inc. and Resilience Shift, United Kingdom.

Zhang H (2021). Chinese Grain Imports: Taking Centre Stage Once Again. Clarksons Research. 30 April.

This chapter reviews the supply of maritime transport, covering the world fleet, shipping companies, and port services, and then adding insights from the UNCTAD TrainForTrade Port Management Programme.

A. The world fleet – This section examines the growth of the world fleet and changes to its structure and age. It also covers parts of the maritime supply chain, such as shipbuilding, ship recycling, ship ownership and ship registration. It finds that at the beginning of 2021 the demand for shipping services was exceeding supply, resulting in a surge in orders for new ships and more activity in the second-hand market.

B. Regulation of shipping – This section examines regulatory changes, in particular decarbonization targets. It explores the implications for the shipping industry and for shipping-related operations, fuel usage and technology. Adapting to these changes will require significant investment at a time of great uncertainty.

C. Port services – This section explains how the pandemic has induced a rethink of business resilience for ports. It also covers their strategies for capitalizing on emerging opportunities, notably ecommerce and greener industrial activities.

D. The impact of COVID-19 – Using data from the UNCTAD TrainForTrade port network, this section examines the impact of the pandemic on financial performance and on vessel and cargo operations.

Maritime transport and infrastructure

Maritime transport services and infrastructure supply

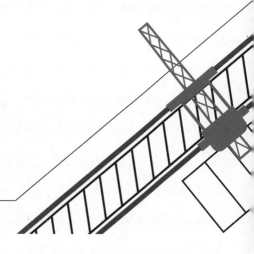

THE WORLD FLEET

In early 2021, the world fleet totalled
99,800 ships
of 100 gross tons and above, equivalent to **2,134,639,907** dwt of capacity

The global shipping fleet grew by
+3%
in the 12 months prior to 1 January 2021

Ships between
5–9 years old
represented the highest proportion of the fleet carrying capacity

Ship deliveries declined by
-12 %
in 2020

SHIPPING COMPANIES AND OPERATIONS

Adapting maritime transport supply

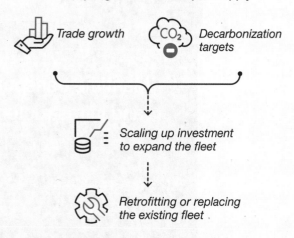

- Trade growth
- Decarbonization targets
- Scaling up investment to expand the fleet
- Retrofitting or replacing the existing fleet

Potential changes from the Green Transition

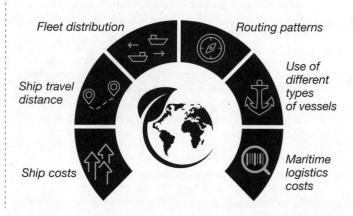

- Fleet distribution
- Routing patterns
- Use of different types of vessels
- Maritime logistics costs
- Ship costs
- Ship travel distance

PORT SERVICES AND INFRASTRUCTURE SUPPLY

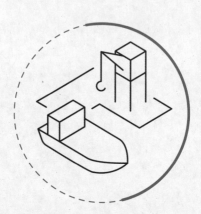

Since 2020, ports resilience and adaptive capacity have been tested:
- Financial performance
- Congestion
- Equipment shortages
- Supply chain disruption

New opportunities from the COVID-19 crisis

E-commerce, smart logistic hubs and intermodal connections

Greener industrial port activities

A. THE WORLD FLEET

1. Fleet structure, age, and vessel size

Ships are getting bigger, though with fewer new ships the fleet is ageing

In the 12 months to 1 January 2021, the global commercial shipping fleet grew by 3 per cent – to 99,800 ships of 100 gross tons and above, equivalent to 2,134,639,907 dwt of capacity (table 2.1). But as indicated in figure 2.1, from a peak of 11 per cent in 2011 this growth rate has slowed.

An increasingly important concern is the ageing of the fleet, since older ships are generally less efficient and generate higher emissions. At the beginning of 2021, around 30 per cent of the carrying capacity of the global fleet was in ships of between five and nine years old (table 2.2). As indicated in figure 2.2, since 2017 this age cohort has represented the highest proportion of capacity, but its proportion and that for younger vessels has been falling, while that for vessels of 10 to 14 years old has steadily been rising.

The age distribution varies, however, between different economies (figure 2.3). The oldest ships are generally those in the least developed countries (LDCs), where close to 30 per cent are more than 20 years old. Compared to the developing group, or the developed countries, the LDCs also have a higher proportion of ships of 15 to 19 years old.

Table 2.1 World fleet by principal vessel type, 2020–2021
(thousand dead-weight tons and percentage change)

Principal types	2020	2021	Percentage change 2021 over 2020
Bulk carriers	879 725	913 032	3.79%
	42.47%	42.77%	
Oil tankers	601 342	619 148	2.96%
	29.03%	29.00%	
Container ships	274 973	281 784	2.48%
	13.27%	13.20%	
Other types of ship:	238 705	243 922	2.19%
	11.52%	11.43%	
Offshore supply	84 049	84 094	0.05%
	4.06%	3.94%	
Gas carriers	73 685	77 455	5.12%
	3.56%	3.63%	
Chemical tankers	47 480	48 858	2.90%
	2.29%	2.29%	
Other/not available	25 500	25 407	-0.36%
	1.23%	1.19%	
Ferries and passenger ships	7 992	8 109	1.46%
	0.39%	0.38%	
General cargo ships	76 893	76 754	-0.18%
	3.71%	3.60%	
World total	**2 071 638**	**2 134 640**	**3.04%**

Source: UNCTAD calculations, based on data from Clarksons Research.

Notes: Propelled seagoing merchant vessels of 100 gross tons and above, at 1 January.
Dead-weight tons for individual vessels have been estimated.

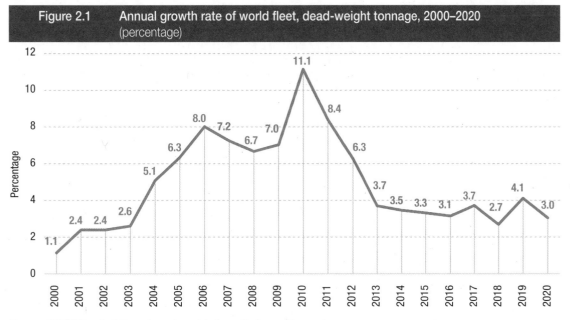

Figure 2.1 Annual growth rate of world fleet, dead-weight tonnage, 2000–2020 (percentage)

Source: UNCTAD calculations, based on data from Clarksons Research.

Table 2.2 Age distribution of world merchant fleet by vessel type, 2021 and average age 2020–2021
(percentage and average vessel size)

Vessel type, country grouping by flag of registration and indicator		Years					Average age	
		0–4	5–9	10–14	15–19	More than 20	2021	2020
World								
Bulk carriers	Percentage of total ships	18	37	24	10	10	10.6	10.2
	Percentage of dead-weight tonnage	22	40	23	9	6	9.5	9.3
	Average vessel size (dead-weight tonnage)	90 447	78 409	68 583	68 087	46 623	NA	NA
Container ships	Percentage of total ships	14	19.21	32	17	17	13.2	12.7
	Percentage of dead-weight tonnage	20	29	29	14	7	10.4	9.9
	Average vessel size (dead-weight tonnage)	74 632	78 802	46 897	42 345	21 975	NA	NA
General cargo	Percentage of total ships	5	10	16	9	59	27.1	26.3
	Percentage of dead-weight tonnage	8	20	23	10	40	19.9	19.3
	Average vessel size (dead-weight tonnage)	5 992	7 493	5 494	4 372	2 660	NA	NA
Oil tankers	Percentage of total ships	14	17	21	13	35	19.5	19
	Percentage of dead-weight tonnage	25	21	28	19	8.	10.9	10.4
	Average vessel size (dead-weight tonnage)	96 122	65 148	72 208	80 802	12 346	NA	NA
Other types of ships	Percentage of total ships	10	17	17	9	47	23.6	23.0
	Percentage of dead-weight tonnage	20	16	23	11	30	16.1	15.8
	Average vessel size (dead-weight tonnage)	9 236	4 562	6 524	5 953	3 014	NA	NA
All ships	Percentage of total ships	11	18	19	10	42	21.6	21.1
	Percentage of dead-weight tonnage	22	29	25	13	11	11.2	10.80
	Average vessel size (dead-weight tonnage)	43 364	34 175	28 112	27 809	5 505	NA	NA
Developing economies (all ships)								
	Percentage of total ships	10	20	19	10	41	20.8	20.2
	Percentage of dead-weight tonnage	21	29	22	13	15	11.9	11.6
	Average vessel size (dead-weight tonnage)	33 788	24 295	18 871	21 144	6 190	NA	NA
Developed economies (all ships)								
	Percentage of total ships	12	17	20	10	40	21.3	20.8
	Percentage of dead-weight tonnage	23	30	28	13	7	10.5	10.2
	Average vessel size (dead-weight tonnage)	54 908	50 000	39 696	35 466	5 132	NA	NA
Small Islands Developing States (all ships)								
	Percentage of total ships	6	8	10	8	68	30.9	30.3
	Percentage of dead-weight tonnage	3	30	18	20	30	17.5	17.8
	Average vessel size (dead-weight tonnage)	2 009	16 865	8 077	11 326	2 036	NA	NA
Least developed countries (all ships)								
	Percentage of total ships	12	13	8	6	61	28.6	28.6
	Percentage of dead-weight tonnage	9	19	25	18	29	17.0	16.5
	Average vessel size (dead-weight tonnage)	7 551	15 032	33 414	31 782	4 956	NA	NA

Source: UNCTAD calculations, based on data from Clarksons Research.

Notes: Propelled seagoing vessels of 100 gross tons and above, as at 1 January.
Dead-weight tons for individual vessels have been estimated.
The LDC and SIDS country grouping are based on the definitions of the Office of the High Representative for the Least Developed Countries, Landlocked Developing Countries and Small Island Developing States (UNOHRLLS). For more information see: https://www.un.org/ohrlls/content/ldc-category and https://www.un.org/ohrlls/content/list-sids.

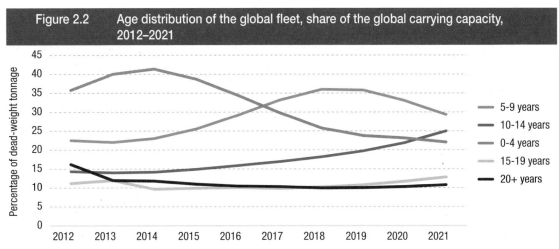

Figure 2.2 Age distribution of the global fleet, share of the global carrying capacity, 2012–2021

Source: UNCTAD calculations, based on data from Clarksons Research.
Notes: Propelled seagoing merchant vessels of 100 gross tons and above; beginning-of-year figures.

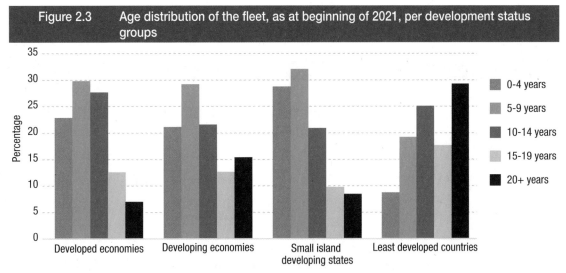

Figure 2.3 Age distribution of the fleet, as at beginning of 2021, per development status groups

Source: UNCTAD calculations, based on data from Clarksons Research.
Note: The LDC and SIDS country grouping are based on the definition by UNOHRLLS. For more information see: https://www.un.org/ohrlls/content/ldc-category and https://www.un.org/ohrlls/content/list-sids.

Increasing ship sizes: what we have learnt from the Ever Given incident

Since the early 2000s, more of the world's cargo has been carried in mega-container ships – those with a container capacity greater than 10,000 twenty-foot equivalent units (TEU): between 2011 and 2021 their proportion of carrying capacity rose from 6 to almost 40 per cent (figure 2.4). In the last 10 years, there have been 97 new ships of between 15,000 and 19,990 TEU, and since 2018 74 ships of 20,000 TEU and above (figure 2.5). These larger ships, facilitated by technological advances, have been part of broader corporate strategies to pursue economies of scale (Sanchez, 2021). However, this has resulted in excess supply – 'over-tonnaging' – in the world's major liner routes, with greater pressure on infrastructure and on logistics at ports.

This pressure on infrastructure was dramatically illustrated from 23 to 29 March 2021 when the Suez Canal was blocked by the Ever Given, a container ship with a carrying capacity of 20,000 TEU. Larger ships are more difficult to steer, and harder and more costly to rescue in cases of collisions and groundings. In addition to safety and salvage issues, the higher risks entail higher insurance costs. (Hayden, 2015; Lockton, 2019; Allianz, 2019; and Boulougouris, 2021).

This is a critical issue for key nodes of the global maritime transport network such as the Suez and Panama canals, which have constrained capacities and where any disruption sends shockwaves through global supply chains. The Ever Given incident delayed the passage of hundreds of vessels through the canal, disrupted global trade, and exacerbated the shortage of shipping containers, leading to congestion in many ports and an increase in container freight rates (Hellenic Shipping News, 2021). As indicated in figure 2.6, since 2012 these mega-vessels have been making more journeys through the Panama and Suez canals.

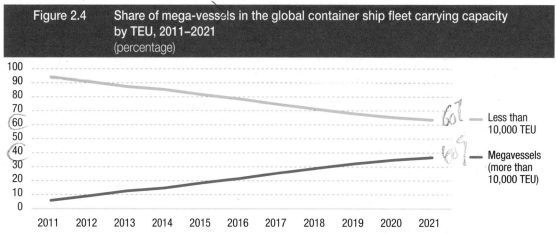

Figure 2.4   Share of mega-vessels in the global container ship fleet carrying capacity by TEU, 2011–2021 (percentage)

Source: UNCTAD calculations, based on data from Clarksons Research.

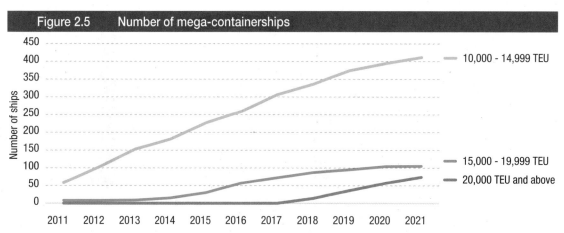

Figure 2.5   Number of mega-containerships

Source: UNCTAD calculations, based on data from Clarksons Research.

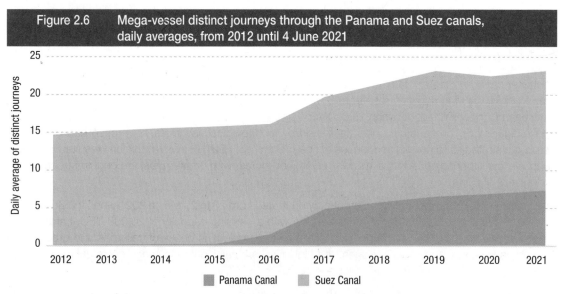

Figure 2.6   Mega-vessel distinct journeys through the Panama and Suez canals, daily averages, from 2012 until 4 June 2021

Source: UNCTAD calculations, based on data from VesselsValue.
Notes: In the case of the Panama Canal the mega-vessel category includes bulk carriers (Capesize), containerships (Neo-Panamax and Post Panamax), gas carriers (Q Flex and VLGC) and oil tankers (VLCC). In the case of the Suez Canal, in addition to the ship types mentioned before, the mega-vessel category includes an additional type of gas carrier (Q-Max) and containership (ULCV).

2. Ship ownership and registration

The principal ship-owning countries mostly flag their ships abroad

As of 1 January 2021, the top three ship-owning countries, in terms of both dead-weight tons and the commercial value of their fleets, were Greece, China, and Japan (table 2.3) (table 2.4). Over the previous year, among the top 35 shipowners, the greatest increases in shares of carrying capacity were in the United Arab Emirates, from 1.01 to 1.18 per cent, and Viet Nam from 0.52 to 0.59 per cent. In terms of value, the highest increases in shares of the world merchant fleet value were in Taiwan Province of China, from 1.49 to 1.86 per cent, and the Republic of Korea, from 2.77 to 3.08 per cent.

Table 2.3 Top 25 ship-owning economies, as of 1 January 2021
(millions of United States dollars)

	Country or Territory of Ownership	Bulk Carriers	Container Ships	Offshore vessels	Oil Tankers	Ferries and Passenger Ships	Gas Carriers	General Cargo Ships	Chemical Tankers	Other/ not available	Total
1	Japan	39 564	15 101	4 746	9 529	3 236	15 436	3 130	5 203	7 888	103 833
2	Greece	39 853	11 670	197	32 602	2 512	14 572	182	977	402	102 968
3	China	34 735	20 632	9 967	12 838	4 979	4 115	5 120	3 344	3 207	98 936
4	United States	3 734	1 938	15 494	5 117	51 259	1 454	1 320	1 098	791	82 206
5	Singapore	14 564	9 274	4 304	12 569	32	4 377	870	4 778	534	51 301
6	Norway	4 384	2 514	21 748	5 570	3 208	7 620	900	2 433	2 719	51 096
7	Germany	6 207	24 166	687	1 767	9 460	1 627	2 789	704	347	47 754
8	United Kingdom	4 001	7 123	10 064	3 829	5 661	5 816	791	1 354	2 239	40 878
9	China, Hong Kong SAR	11 117	12 982	73	6 288	2 387	1 114	918	269	886	36 032
10	Republic of Korea	9 123	5 363	240	5 558	433	4 791	680	1 480	2 673	30 340
11	Bermuda	5 863	2 301	5 198	5 919		8 107		297	51	27 736
12	Denmark	1 526	12 847	1 701	3 416	1 032	2 049	751	1 032	108	24 462
13	Switzerland	822	9 012	3 056	596	9 521	213	183	169	12	23 584
14	Netherlands	704	412	13 273	441	526	686	2 969	1 892	2 046	22 949
15	Taiwan Province of China	8 145	7 372	48	1 483	74	363	563	148	107	18 304
16	Italy	1 116	6	2 441	1 866	9 475	256	1 801	418	621	18 000
17	Brazil	179	465	14 312	810	64	116	30	77	2	16 054
18	Monaco	3 390	2 004		6 381	29	3 300		26	24	15 153
19	France	374	5 325	5 183	112	1 860	476	155	132	144	13 761
20	Russian Federation	256	110	1 346	3 320	76	1 740	1 449	637	1 828	10 762
21	Turkey	3 406	1 011	677	1 269	353	131	1 793	1 156	51	9 847
22	Indonesia	1 110	1 103	1 137	2 131	2 020	565	1 174	369	51	9 659
23	Malaysia	142	110	6 748	219	19	1 811	189	150	159	9 548
24	Belgium	1 747	491	134	3 305		860	761	210	2 018	9 526
25	United Arab Emirates	1 959	469	2 858	2 361	57	544	90	621	179	9 138
	Others	14 436	4 971	23 462	18 470	12 008	13 971	7 863	4 050	2 297	101 529
	World total	212 455	158 771	149 093	147 764	120 282	96 110	36 470	33 026	31 384	985 356

Source: UNCTAD calculations, based on data from Clarksons Research, as of 1 January 2021 (estimated current value).
Note: Value is estimated for all commercial ships of 1,000 gross tons and above.

Table 2.4 Ownership of the world fleet, ranked by carrying capacity in dead-weight tons, 2021

	Country or territory of ownership	Number of vessels — National flag	Number of vessels — Foreign flag	Number of vessels — Total	Deadweight tonnage — National flag	Deadweight tonnage — Foreign flag	Deadweight tonnage — Total	Foreign flag as a percentage of total	Total as a percentage of world
1	Greece	642	4 063	4 705	58 067 003	315 350 152	373 417 155	84.45%	17.64%
2	China	4 887	2 431	7 318	105 657 323	138 898 420	244 555 743	56.80%	11.56%
3	Japan	914	3 115	4 029	35 107 223	206 741 103	241 848 326	85.48%	11.43%
4	Singapore	1 459	1 384	2 843	73 258 302	65 805 758	139 064 059	47.32%	6.57%
5	China, Hong Kong SAR	886	878	1 764	72 367 151	31 851 549	104 218 700	30.56%	4.92%
6	Germany	198	2 197	2 395	7 437 473	78 759 307	86 196 779	91.37%	4.07%
7	Republic of Korea	787	854	1 641	15 096 916	70 995 920	86 092 836	82.46%	4.07%
8	Norway	387	1 655	2 042	1 899 017	62 144 480	64 043 497	97.03%	3.03%
9	Bermuda	13	540	553	300 925	63 733 226	64 034 151	99.53%	3.03%
10	United Kingdom (excl. Channel Islands)	309	1 014	1 323	7 160 493	46 524 174	53 684 667	86.66%	2.54%
11	United States of America (incl. Puerto Rico but excluding Virgin Islands)	790	1 020	1 810	10 395 172	44 576 019	54 971 191	81.09%	2.60%
12	Taiwan Province of China	147	867	1 014	6 998 235	46 284 542	53 282 777	86.87%	2.52%
13	Monaco	0	478	478	0	43 426 478	43 426 478	100.00%	2.05%
14	Denmark	26	902	928	47 415	42 185 673	42 233 088	99.89%	2.00%
15	Belgium	108	249	357	8 974 783	21 969 171	30 943 954	71.00%	1.46%
16	Turkey	429	1 112	1 541	5 994 812	21 970 706	27 965 518	78.56%	1.32%
17	Indonesia	2 232	89	2 321	24 139 035	2 704 715	26 843 751	10.08%	1.27%
18	Switzerland	18	396	414	928 432	25 794 797	26 723 229	96.53%	1.26%
19	India	875	195	1 070	16 396 087	10 013 434	26 409 521	37.92%	1.25%
20	United Arab Emirates	119	941	1 060	525 959	24 431 420	24 957 380	97.89%	1.18%
21	Russian Federation	1 464	322	1 786	9 184 626	14 682 694	23 867 320	61.52%	1.13%
22	Iran (Islamic Republic of)	246	8	254	18 898 257	352 889	19 251 146	1.83%	0.91%
23	Netherlands	692	515	1 207	5 577 088	13 185 003	18 762 090	70.27%	0.89%
24	Saudi Arabia	151	111	262	13 397 363	3 422 203	16 819 566	20.35%	0.79%
25	Italy	481	170	651	10 296 714	5 900 509	16 197 223	36.43%	0.77%
26	Brazil	292	91	383	4 735 593	9 120 015	13 855 608	65.82%	0.65%
27	France, metropolitan	98	327	425	1 592 919	12 004 098	13 597 017	88.28%	0.64%
28	Viet Nam	929	166	1 095	9 491 311	3 043 458	12 534 769	24.28%	0.59%
29	Cyprus	134	177	311	5 166 089	7 174 723	12 340 812	58.14%	0.58%
30	Canada	210	164	374	2 569 373	7 212 024	9 781 397	73.73%	0.46%
31	Oman	5	58	63	5 704	8 926 419	8 932 123	99.94%	0.42%
32	Malaysia	456	163	619	6 587 734	2 158 859	8 746 592	24.68%	0.41%
33	Qatar	57	69	126	1 123 717	6 145 431	7 269 149	84.54%	0.34%
34	Nigeria	198	73	271	3 517 645	3 429 887	6 947 532	49.37%	0.33%
35	Sweden	90	208	298	1 004 333	5 448 524	6 452 857	84.44%	0.30%
	Subtotal, top 35 shipowners	20 729	27 002	47 731	543 900 223	1 466 373 485	2 010 273 707	72.94%	94.99%
	Rest of the world unknown	3 096	3 146	6 242	37 011 088	69 116 093	106 127 181	65.13%	5.01%
	World	23 825	30 148	53 973	580 911 310	1 535 489 578	2 116 400 888	72.55%	100.00%

Source: UNCTAD calculations, based on data from Clarksons Research.

Notes: Propelled seagoing vessels of 1,000 gross tons and above, as of 1 January 2021. For the purposes of this table, second and international registries are recorded as foreign or international registries, whereby, for example, ships belonging to owners in the United Kingdom but registered in Gibraltar or on the Isle of Man are recorded as being under a foreign or international flag. In addition, ships belonging to owners in Denmark and registered in the Danish International Ship Register account for 48 per cent of the Denmark-owned fleet in dead-weight tonnage, and ships belonging to owners in Norway registered in the Norwegian International Ship Register account for 28 per cent of the Norway-owned fleet in dead-weight tonnage.
For a complete listing of nationally owned fleets, see http://stats.unctad.org/fleetownership.

2. Maritime transport and infrastructure

Rising value of the fleet: a sign of confidence?

The commercial value of a vessels depends on many considerations, including: size, type, builder, age, classification status, certifications, ship condition and maintenance, added technology, and engine and fuel efficiency. Values are also influenced by prevailing conditions in shipping and financial markets. As of 1 June 2021, the highest value was in bulk carriers at 27 per cent, followed by container ships at 25 per cent, and tankers at 22 per cent (figure 2.7). For ships on order, the highest value was in container ships 30 per cent, followed by tankers at 20 per cent and LNG carriers at 16 per cent.

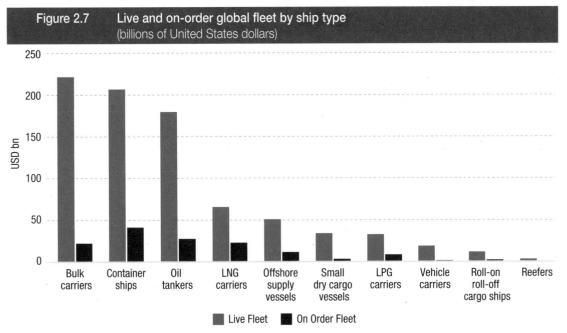

Figure 2.7 Live and on-order global fleet by ship type
(billions of United States dollars)

Source: UNCTAD, based on data from VesselsValue, as of 1 June 2021.
Note: Includes all vessels above 1,000 GT.

Second-hand ship prices can be quite volatile. Since the last quarter of 2020, there have, for example, been significant increases in the value of container ships. Between end-2020 and mid-June 2021 the Containership Secondhand Price Index increased by 71 per cent. Sales were at their highest since 2013, reflecting the demand for smaller container ships of between 5 and 15 years old (Clarksons Research, 2021a).

There have also been significant increases in the prices for second-hand bulk carriers. Since October 2020, the Bulk Carrier Secondhand Price Index has been steadily increasing – during the first half of 2021 prices of various vessel sizes aged between 5 and 10 years rose by between 25 and 50 per cent (Miller, 2021). Higher prices reflect strong short-term market confidence, based on rising commodity prices, high earnings for bulk carriers and projections for increasing global seaborne bulk trade (Clarksons Research, 2021b). Since the beginning of 2021, sales have been at their highest for the past five years (Roussanoglou, 2021a).

To a great extent, selling and purchasing decisions are driven by expected future profitability (Haralambides et al. 2005). In times of tight vessel supply, higher freight rates drive up the prices of ships (see chapter 3). The stronger market for used vessels may also signal a return in investor confidence. By buying second-hand ships, companies can expand rapidly by acquiring almost instantly available tonnage (Sancricca, 2016).

Developing economies remain the main providers of ship registration

As of 1 January 2021, in terms of both carrying capacity (table 2.5) and commercial value of the fleet, the top three flags of registration remained those of Panama, Liberia and Marshall Islands (table 2.6). Among the top 35 flags of registration, the greatest increases were in Viet Nam by 12.1 per cent, from 9,868 to 10,269 thousand dwt, and in the Russian Federation, by 10.4 per cent, from 9,164 to 10,899 thousand dwt. In terms of value, the greatest increase was in Nigeria whose share of the world merchant fleet value increased from 0.50 to a 0.78 per cent.

Table 2.5 Leading flags of registration by dead-weight tonnage, 2021

	Flag of registration	Number of vessels	Share of world vessel total (percentage)	Dead-weight tonnage (thousands dead-weight tons)	Share of total world dead-weight tonnage (percentage)	Cumulative share of dead-weight tonnage (percentage)	Average vessel size (dead-weight tonnage)	Growth in dead-weight tonnage 2020 to 2021
1	Panama	7 980	8	344 200	16.1	16.1	43 133	4.6
2	Liberia	3 942	4	300 088	14.1	30.2	76 126	8.9
3	Marshall Islands	3 817	4	274 041	12.8	43.0	71 795	4.7
4	Hong Kong, China	2 718	3	205 092	9.6	52.6	75 457	1.8
5	Singapore	3 321	3	136 400	6.4	59.0	41 072	-2.6
6	Malta	2 137	2	116 407	5.5	64.5	54 472	0.5
7	China	6 653	7	107 583	5.0	69.5	16 171	5.0
8	Bahamas	1 323	1	74 289	3.5	73.0	56 152	-4.3
9	Greece	1 236	1	64 850	3.0	76.0	52 468	-6.0
10	Japan	5 201	5	39 091	1.8	77.9	7 516	-3.6
11	Cyprus	1 051	1	33 976	1.6	79.5	32 328	-1.6
12	Indonesia	10 427	10	28 750	1.3	80.8	2 757	6.0
13	Danish International Register	602	1	24 735	1.2	82.0	41 089	6.9
14	Madeira	578	1	22 726	1.1	83.0	39 318	9.7
15	Norwegian Int'l Register	671	1	22 093	1.0	84.1	32 926	5.7
16	Isle of Man	319	0	22 011	1.0	85.1	68 999	-8.7
17	Iran (Islamic Republic of)	893	1	20 417	1.0	86.0	22 863	3.1
18	India	1 801	2	17 054	0.8	86.8	9 469	-2.1
19	Republic of Korea	1 904	2	15 723	0.7	87.6	8 258	4.9
20	Saudi Arabia	392	0	13 662	0.6	88.2	34 853	-1.7
21	United States	3 625	4	12 456	0.6	88.8	3 436	-0.4
22	United Kingdom	927	1	12 063	0.6	89.4	13 013	-0.2
23	Italy	1 296	1	11 255	0.5	89.9	8 685	-6.1
24	Russian Federation	2 873	3	10 899	0.5	90.4	3 794	10.4
25	Viet Nam	1 926	2	10 269	0.5	90.9	5 332	12.1
26	Malaysia	1 769	2	10 231	0.5	91.4	5 783	-1.6
27	Belgium	201	0	9 603	0.4	91.8	47 774	-4.5
28	Bermuda	147	0	8 053	0.4	92.2	54 781	3.0
29	Germany	598	1	7 618	0.4	92.6	12 740	-10.7
30	Taiwan Province of China	429	0	7 136	0.3	92.9	16 635	5.3
31	Netherlands	1 199	1	6 807	0.3	93.2	5 677	-3.4
32	Cayman Islands	160	0	6 725	0.3	93.5	42 032	0.1
33	Turkey	1 217	1	6 425	0.3	93.8	5 279	-9.2
34	Antigua and Barbuda	677	1	6 402	0.3	94.1	9 456	-3.5
35	Philippines	1 805	2	6 240	0.3	94.4	3 457	-5.3
	Top 35	75 815	76	2 015 370	94.4	94.4	26 583	2.7
	World total	99 800	100	2 134 640	100.0	100.0	21 389	3.0

Source: UNCTAD calculations, based on data from Clarksons Research.

Notes: Propelled seagoing merchant vessels of 100 gross tons and above, as of 1 January 2021. For a complete listing of countries, see http://stats.unctad.org/fleet.
Dead-weight tons for individual vessels have been estimated.

Table 2.6 Leading flags of registration, ranked by value of total tonnage, 2021 (million US dollars) and principal vessel types

	Flag of Registration	Bulk carriers	Container ships	Offshore vessels	Oil tankers	Ferries and passenger ships	Gas carriers	General cargo ships	Chemical tankers	Other/ not applicable	Total
1	Panama	46 903	23 289	14 056	12 065	12 786	10 108	3 768	5 260	6 314	134 550
2	Marshall Islands	32 671	8 217	12 787	26 845	1 513	14 537	430	4 470	1 917	103 388
3	Liberia	29 781	26 351	10 520	20 941	430	5 977	796	2 862	1 439	99 097
4	Bahamas	5 177	706	22 781	6 521	28 250	12 000	65	74	2 303	77 878
5	Hong Kong, China	25 050	25 442	260	10 404	42	6 439	1 318	1 687	105	70 747
6	Malta	10 205	14 925	4 240	9 448	15 166	6 407	1 740	1 661	834	64 626
7	Singapore	13 509	16 531	7 589	11 445		7 947	803	3 560	1 189	62 571
8	China	16 555	5 609	7 728	8 023	4 159	731	2 885	1 668	3 079	50 436
9	Italy	650	196	284	852	15 027	200	1 826	327	621	19 985
10	Greece	3 305	245	1	8 375	1 338	5 388	52	82	22	18 808
	Subtotal top 10	183 806	121 512	80 246	114 918	78 711	69 735	13 684	21 651	17 823	702 087
	Other	28 649	37 260	68 847	32 846	41 571	26 375	22 785	11 375	13 561	283 269
	World total	212 455	158 771	149 093	147 764	120 282	96 110	36 470	33 026	31 384	985 356

Source: UNCTAD calculations, based on data from Clarksons Research, as at 1 January 2019 (estimated current value).
Note: Value is estimated for all commercial ships of 1,000 gross tons and above.

3. Shipbuilding, new orders and ship recycling

Two-thirds of world ship building was of dry bulk carriers and tankers

In 2020, ship deliveries declined by 12 per cent, mainly due to lockdown-induced labour shortages during the first half of the year that disrupted marine-industrial activity. As in 2018 and 2019, the ships delivered were mostly bulk carriers, followed by oil tankers and container ships (table 2.7). Since 2015, an increasing proportion of shipbuilding has taken place in just four countries – China, the Republic of Korea, Japan, and the Philippines. In 2020, their combined market share rose to 96 per cent.

Table 2.7 Deliveries of newbuildings by major vessel types and countries of construction, 2020 (thousand gross tons)

Vessel type	China	Republic of Korea	Japan	Philippines	Rest of the world	Total	Percentage
Bulk carriers	15 051	1 442	9 383	551	311	**26 738**	46
Oil tankers	2 702	7 071	1 901	1	478	**12 152**	21
Container ships	2 665	5 357	394	56	200	**8 671**	15
Gas carriers	869	4 046	353		7	**5 275**	9
Ferries and passenger ships	251	64	76		1 208	**1 600**	3
Chemical tankers	488	88	465		55	**1 095**	2
General cargo	390	1	142		360	**893**	2
Offshore	340	101	7		118	**566**	1
Other	501	4	107		162	**775**	1
Total	23 257	18 174	12 827	608	2 898	57 765	100
Percentage	*40*	*31*	*22*	*1*	*5*	*100*	

Source: UNCTAD calculations, based on data from Clarksons Research.
Notes: Propelled seagoing merchant vessels of 100 gross tons and above. For more data on other shipbuilding countries, see http://stats.unctad.org/shipbuilding.

China has the largest share at around 40 per cent. Since the 1980s, based on cost advantages and with strong government policy support, China's shipbuilding industry has sought to improve its capabilities and expand capacity. In 1982, the shipbuilding ministry was 'corporatized' as the China State Shipbuilding Corporation (CSSC) which now administers most commercial and military shipbuilding. This prioritized development in prosperous coastal regions through decentralized organization of diverse related industries. Focussing on international demand, the industry also had greater access to foreign capital, and in the last two decades Chinese companies have entered into technology-sharing agreements with foreign shipbuilders giving them access to foreign equipment, materials and technical expertise. R&D institutes and academic organizations in China have also enhanced their research, development and design capabilities (Market and Research News, 2021 and Medeiros et al. 2021). As a result, over recent years China has improved its building techniques and efficiency and increased its market share not just for bulk carriers and container ships but also for segments where it has previously not operated, such as passenger ships and LNG carriers (Hellenic Shipping News, 2021).

New orders

Between January 2020 and January 2021, the global orderbook declined by 16 per cent. The sharpest reductions were for bulk carriers, down 36 per cent, followed by ferries and passenger ships, down 32 per cent. By contrast, other segments grew: liquefied gas carriers, up 10 per cent, and general cargo ships, up 6 per cent (figure 2.8).

From a longer-term perspective, the fleet orderbook has been shrinking since 2011, reaching 165,520,744 dwt in January 2021, the lowest level for the last decade. This is largely the result of constraints on finance combined with uncertainty over future choices of energy sources, and compounded from 2020 by the impacts of COVID-19 on trade volumes and economic activity. At the beginning of 2021, order levels for container ships were similar to those in 2018, for bulk carriers to those in 2004–2006, and for oil tankers to those in 2001, 2003 and 2020 (figure 2.9).

Since early 2021, however, there has been a surge of new orders. As world trade gradually recovered during the second half of 2020 and the first half of 2021, demand for ships increased – responding to severe fleet capacity constraints and the uptick in freight rates. In the first half of 2021, newbuild investment was at its highest since the first half of 2014 (Bak, 2021), with record-breaking orders for container ships – almost eight times those in the first half of 2020. New building orders were spearheaded by those for Panamax container ships (ShipInsights, 2021). There has also been an increase for LNG carriers (Roussanoglou, 2021b).

The largest increases in orders during this period were for Chinese and Korean shipbuilders (Maritime Executive, 2021). However, these orders appear to be concentrated in a few shipyards – which could increase average contract lead times and hinder fleet growth (Springer, 2021, and Walia, 2021).

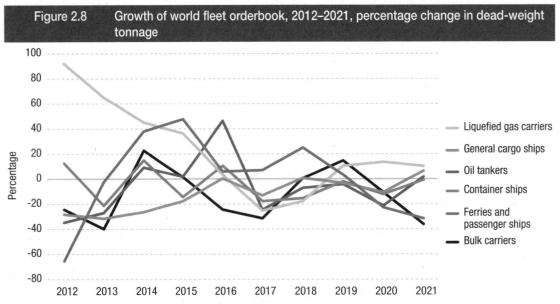

Figure 2.8   Growth of world fleet orderbook, 2012–2021, percentage change in dead-weight tonnage

Source: UNCTAD calculations, based on data from Clarksons Research.
Notes: Propelled seagoing merchant vessels of 100 gross tons and above; beginning-of-year figures.

2. Maritime transport and infrastructure

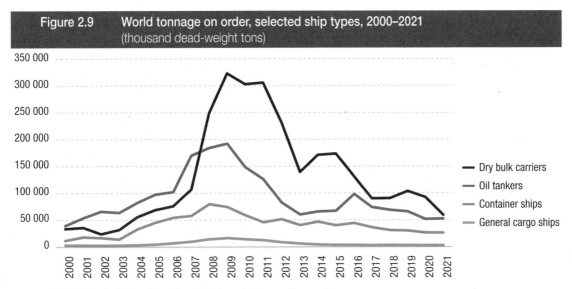

Figure 2.9 World tonnage on order, selected ship types, 2000–2021
(thousand dead-weight tons)

Source: UNCTAD calculations, based on data from Clarksons Research.
Notes: Propelled seagoing merchant vessels of 100 gross tons and above; beginning-of-year figures.

Ship recycling

Even through the COVID-19 disruption, the tonnage of ships sold for recycling increased by 44 per cent in 2020, reaching 17,400,564 GT. Nevertheless, recycling levels remain lower than in the 2014–2017 period. Despite high scrap metal prices, ship owners believe they can continue to earn high incomes by continuing to operate older vessels.

In 2020, almost half of the recycling was of bulk carriers, reflecting declining charter rates and following the trend of recycling ageing tonnage (Jiang, 2021 and Clarksons Research, 2021c). Around two-thirds of reported tonnage sold for recycling in 2020 was in Bangladesh and India. With the addition of Pakistan and Turkey, the share of the top four countries reached 93 per cent (table 2.8). The highest increases in shares were for Pakistan, by 14.7 percentage points, and for India by 3.2 percentage points.

In contrast, there were noticeable reductions in Bangladesh, by 15 percentage points, and in China by 2 percentage points. In China, this follows a ban on receiving international vessels for recycling, which

Table 2.8 Reported tonnage sold for ship recycling by major vessel type and country of ship recycling, 2020
(thousand gross tons)

Vessel type	Bangladesh	India	Pakistan	Turkey	China	Rest of the world	World total	Percentage
Bulk carriers	5 254	1 317	1 718	34	125	61	**8 509**	48.9
Container ships	160	1 428	282	206		68	**2 143**	12.3
Oil tankers	616	410	617	159	10	226	**2 038**	11.7
Offshore supply	125	257	4	308	3	273	**969**	5.6
Ferries and passenger ships	26	279		545	3	26	**879**	5.1
General cargo ships	176	219	175	203	47	29	**848**	4.9
Liquefied gas carriers	169	241		8		176	**594**	3.4
Chemical tankers	12	125	94	1		10	**241**	1.4
Other/ n.a.	157	786		135	9	93	**1 180**	6.8
Total	**6 694**	**5 061**	**2 890**	**1 598**	**195**	**962**	**17 401**	**100.0**
Percentage	*38.5*	*29.1*	*16.6*	*9.2*	*1.1*	*5.5*	*100.0*	

Source: UNCTAD calculations, based on data from Clarksons Research.
Notes: Propelled seagoing vessels of 100 gross tons and above. Estimates for all countries available at http://stats.unctad.org/shiprecycling.

entered into force in 2018. Between 2017 and 2020, China's share of global recycling tonnage fell from 16 to 1 per cent.

The extent of ship recycling depends on a number of factors, including vessel age, freight markets, and trade patterns (OECD, 2019). In addition, ship owners have to take into account new environment-related regulations, such as IMO limits on the sulphur content of ship fuel oil, the IMO Ballast Water Management Convention, and emerging IMO regulations on decarbonization. When capital expenditures for retro-fitting older ships to comply with new regulations exceed the return on investment, owners are likely to favour recycling.

B. SHIPPING COMPANIES AND OPERATIONS: ADAPTING MARITIME TRANSPORT SUPPLY IN AN UNCERTAIN ENVIRONMENT

1. Expanding and renewing the global fleet

Until recently, there was a structural oversupply of maritime transport and, especially from the onset of the pandemic, ship owners had been cutting capacity. Since 2021, however, supply has lagged behind demand, leading to higher freight rates (UNCTAD, 2021a).

This situation poses fundamental questions about the future of maritime transport. Owners now have to decide what ships they require to expand and renew their fleets, and must do so in an uncertain environment. This also means taking into account significant regulatory changes, particularly those related to decarbonization and the aim of zero emissions (Shell and Deloitte, 2020). To achieve this, the industry needs to consider measures and technologies that can improve ship efficiency. These include:

- Lightweight materials
- Slender hull design
- Propulsion improvement
- Bulbous bows
- Air lubrication systems
- Advanced hull coating
- Ballast water-system design
- Engine and auxiliary systems improvement
- Higher efficiency standards

Some of these options are being incorporated in newbuilds or in the orderbook but, as indicated in table 2.9, they have yet to be widely deployed in the global fleet. Others are not yet economically viable (Balcombe et al. 2019).

Table 2.9	Status of uptake of selected technologies in global shipping, as of 14 June 2021		
Equipment type	Energy-saving technologies	Ballast water management systems	(Modern) eco-engine
Fleet, number of ships	3 929	18 925	6 698
Percent of fleet (Percent of GT capacity)	3.9% (19.0%)	18.8% (59.5%)	6.7% (25.7%)
Orderbook	254	2 078	
Percent of orderbook (Percent of GT capacity)	6.8% (13.2%)	55.3% (91.6%)	

Source: Clarksons Research (2021). Tracking "Green" Technology Uptake - June 2021 and Eco-fleet dashboard. *Shipping Intelligence Network.*

Notes: As of 14th June 2021, the global fleet (vessels above 100GT) stood at 100,500 ships, as per Clarksons data. Energy-saving technologies encompass waste heat recovery systems, exhaust gas economizers, propeller ducts, pre-Swirl or stator fins, rudder bulbs, rigid sails, air lubrication system, bow enhancement and solar panels. Modern eco-engine refers to a vessel with an electronic injection main engine contracted after 1st January 2012.
Data based on reported equipment in merchant fleet, which may underestimate total uptake.

Responding to this challenge will require significant investment. Expanding the fleet to cater for trade growth over the coming three decades could cost around $0.2 trillion while retrofitting or replacing the existing fleet over the next 30 years, could cost an additional $2.19 trillion (Ovcina, 2021).[1] Since it is impossible to renew the whole fleet by 2050, innovation and new technologies will also need to be applied to existing vessels.

2. Decarbonization without a crystal ball

Uncertain decarbonization scenarios

In 2018, the IMO adopted a sector reduction pathway consistent with the Paris Agreement. The aim is by 2050 to reduce total annual greenhouse gas emissions by at least 50 per cent of 2008 levels, while reducing carbon intensity by at least 40 per cent by 2030, and 70 per cent by 2050. These objectives are to be achieved through a combination of short-, mid- and long-term measures, with quantitative targets until 2050. Table 2.10 summarizes some proposed measures.

Table 2.10 Some proposed IMO measures to reduce greenhouse gas emissions

Category	Subcategories	Examples of measures
Short-term measures, to be agreed upon between 2018 and 2023	• Technical and operational energy-efficiency measures • Use of alternative low-carbon or zero-carbon fuels for marine propulsion and other technologies	• New operational energy-efficiency standards for new and existing ships (EEXI) • Consider and analyse the use of speed optimization and reduction • Developments of port infrastructure to support alternative fuels • Progressive tightening of standards on minimum energy efficiency levels and emissions, based on ship design and engine performance data (CII) • R&D efforts on marine propulsion with alternative fuels • Encourage the development of national action plans to develop policies and strategies to address greenhouse gas emissions from international shipping
Mid-term measures, to be agreed upon between 2023 and 2030	• Market-based measures – carbon pricing mechanisms to give firms economic incentives to emit less • Operational energy efficiency measures for new and existing ships	• Market-based measures could include an offsetting scheme, a maritime emissions trading scheme, or a carbon levy • Specify in the national action plan measures to increase the uptake of low- and zero-carbon fuels
Long-term measures (to be agreed beyond 2030)	• Measures to ensure zero-carbon and fossil-free fuels	

Sources: IMO (2018), Kachi et al. (2019).

Note: Some measures mentioned in this table have been agreed at the IMO (short-term measures including EEXI and CII) whereas others have not.

At present, the regulatory outlook is uncertain. The IMO has yet to agree on a number of issues, such as the market-based mechanism, and the outcome is likely to be combination of measures. Moreover, the IMO regulations will be accompanied by those from other bodies such as the EU. On 14 July 2021 the EU announced a series of measures:

- Including ships of 5,000 GT and above in its Emissions Trading System for all intra-EEA voyages and for 50 per cent of voyages starting and ending in the bloc.
- Establishing greenhouse gas intensity standards for ship fuels.
- Introducing taxes on bunkers sold in the European Economic Area.

The interplay between different regulatory regimes, combined with volatility in carbon prices is generating considerable uncertainty – which is compounded by the difficulty in modelling the outcome of each measure (ING, 2021). Total emissions will depend on ship type, size and engine, as well as on sea routes

---

[1] These projections exclude fuel transition-related investments, such as storage and transport of alternative fuels.

and navigation conditions – information which may not be easily accessible (Sanchez et al, 2020 and Plevrakis, 2020).

Since 2020, UNCTAD has been collaborating with IMO on assessing the impact of short-term measures. In a report published in 2021, UNCTAD looks at the combined impact of two measures: a new energy standard, the energy efficiency existing ship index; and a new operational requirement, the carbon intensity indicator (UNCTAD, 2021b). The report considers their potential impacts on ship costs, travel distances, fleet distribution, routing patterns, and the use of different types of vessels as well as on maritime logistics costs. The report concludes that the greatest impact will be on smaller vessels plying shorter routes and on container ships and tanker vessels (figure 2.10).

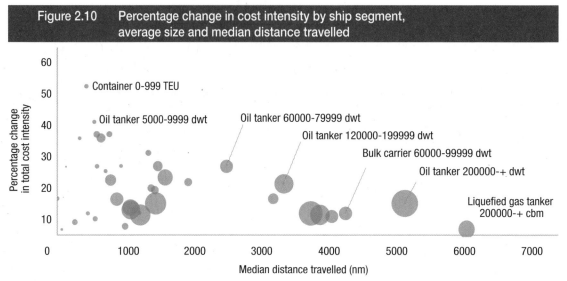

Figure 2.10 Percentage change in cost intensity by ship segment, average size and median distance travelled

Source: UNCTAD compiled from DNV and MarineTraffic data.

Notes: Size of the bubbles stands for the average ship size per DWT. This figure represents the percentage change in total cost intensity between (i) the most ambitious greenhouse gas reduction scenario (regulatory scenario including both EEXI and CII requirements, with an average CII reduction requirement of 21.5 per cent between 2019 and 2030) and (ii) the 2030 "current regulations scenario (with only adopted EEDI requirements, including those entering into force in 2022)".

The easiest and cheapest way to reduce emissions is to reduce ship speed. Operating at less than full power cuts fuel consumption, and thus carbon emissions, while reducing operating costs. However, transporting the same cargo volumes at slower speeds will also require more ships. The report estimates that the IMO short-term measures will require 13 per cent more vessel capacity. This will entail considerable capital expenditure and have important implications for shipbuilders. Drewry estimates that global shipbuilding capacity is equivalent to 7 per cent of the global fleet and that, while maintaining also normal fleet replacement and growth, increasing vessel capacity by 13 per cent would require a ramp-up period of around five years (UNCTAD, 2021b).

The study points out that reducing speeds will also mean reconfiguring services – especially for Pacific and Caribbean SIDS where the maritime trade typically depends on smaller cargo ships on shorter routes. Smaller ships will also be needed, when a deep-sea liner that is going slower now needs to skip a port – which would require more transhipment, thereby increasing costs.

Uncertain energy transition pathways

The path towards shipping decarbonization involves not just ship design and improvements in technology but also the use of alternative fuels. As indicated in table 2.11, the shipping industry uses a range of fuels, though the predominant ones are traditional liquid ones, such as very-low-sulphur intermediate fuel oil (VLS IFO) and intermediate fuel oil with a maximum viscosity of 380 centistokes (IFO380), along with VLS marine diesel oil.

There is certainly significant scope for moving the existing fleet to alternative fuels but there are many areas of uncertainty, and the shift to net-zero fuels has barely begun. For alternative fuels it is important to ensure their safety and consider upstream emissions from their production (see box 2.1).

2. Maritime transport and infrastructure

Table 2.11 World fleet by fuel type as of 1 January 2021

Fuel type	Ships	GT	TEU	dwt	Ships %	GT %	TEU %	Dwt %	Ships % of known fuel type	GT % of known fuel type	TEU % of known fuel type	Dwt % of known fuel type
Very Low-Sulphur (VLS) Intermediate Fuel Oil (IFO)	36 188	993 715 259	18 384 210	1 534 083 046	36.26	69.08	70.97	72.11	47.12	72.26	71.29	74.54
VLS Marine Diesel Oil (MDO)	33 118	29 698 675	149 929	27 886 341	33.18	2.06	0.58	1.31	43.12	2.16	0.58	1.36
IFO 380*	3 635	283 299 533	6 949 482	437 386 040	3.64	19.69	26.83	20.56	4.73	20.60	26.95	21.25
VLS Marine Gasoil (MGO)	2 539	7 441 142	34 467	6 769 951	2.54	0.52	0.13	0.32	3.31	0.54	0.13	0.33
Ultra-Low Sulphur (ULS) MDO	381	697 587	7 000	661 627	0.38	0.05	0.03	0.03	0.50	0.05	0.03	0.03
LNG, VLS IFO	373	36 964 811	144 014	30 159 817	0.37	2.57	0.56	1.42	0.49	2.69	0.56	1.47
LNG, VLS MDO	168	10 814 060	12 703	8 190 743	0.17	0.75	0.05	0.39	0.22	0.79	0.05	0.40
IFO 180	166	7 351 589	75 955	9 536 173	0.17	0.51	0.29	0.45	0.22	0.53	0.29	0.46
ULS IFO	43	352 580	15 617	438 639	0.04	0.02	0.06	0.02	0.06	0.03	0.06	0.02
LNG, VLS MGO	37	424 846	10	430 662	0.04	0.03	0.00	0.02	0.05	0.03	0.00	0.02
LNG	32	459 380	260	139 039	0.03	0.03	0.00	0.01	0.04	0.03	0.00	0.01
MDO	22	652 797	1 629	188 652	0.02	0.05	0.01	0.01	0.03	0.05	0.01	0.01
ULS MGO	22	26 594		16 571	0.02	0.00		0.00	0.03	0.00		0.00
Biofuel	18	360 677	11 684	386 434	0.02	0.03	0.05	0.02	0.02	0.03	0.05	0.02
MGO	12	880 222		122 003	0.01	0.06		0.01	0.02	0.06		0.01
Methanol, VLS IFO	11	336 377		552 044	0.01	0.02		0.03	0.01	0.02		0.03
Ethane, VLS IFO	7	292 595		264 750	0.01	0.02		0.01	0.01	0.02		0.01
Nuclear	6	144 573	1 324	50 079	0.01	0.01	0.01	0.00	0.01	0.01	0.01	0.00
LPG, VLS IFO	5	236 752		272 690	0.01	0.02		0.01	0.01	0.02		0.01
Biofuel, LNG	4	43 851		3 907	0.00	0.00		0.00	0.01	0.00		0.00
Compressed Natural Gas (CNG), VLS MDO	3	111 058		105 325	0.00	0.01		0.00	0.00	0.01		0.01
IFO 380, LNG	2	251 144		18 400	0.00	0.02		0.00	0.00	0.02		0.00
MDO, MGO	2	183 254		16 030	0.00	0.01		0.00	0.00	0.01		0.00
Biofuel, VLS MGO	2	6 810		9 876	0.00	0.00		0.00	0.00	0.00		0.00
VLS IFO, Well Fuel	1	86 952		166 546	0.00	0.01		0.01	0.00	0.01		0.01
CNG, VLS MGO	1	30 742		31 473	0.00	0.00		0.00	0.00	0.00		0.00
LNG, MDO	1	65 314	600	22 437	0.00	0.00	0.00	0.00	0.00	0.00	0.00	0.00
IFO 380*, MGO	1	149 215		19 189	0.00	0.01		0.00	0.00	0.01		0.00
Methanol	1	51 837		10 670	0.00	0.00		0.00	0.00	0.00		0.00
Nuclear, VLS MDO	1	33 500		9 000	0.00	0.00	-	0.00	0.00	0.00	-	0.00
Unknown fuel type	22 998	63 435 988	115 238	69 356 421	23.04	4.41	0.44	3.26				
Grand Total	**99 800**	**1 438 599 714**	**25 904 122**	**2 127 304 575**	**100.00**	**100.00**	**100.00**	**100.00**				
World total known fuel type	**76 802**	**1 375 163 726**	**25 788 884**	**2 057 948 154**					**100.00**	**100.00**	**100.00**	**100.00**

Source: UNCTAD, based on data provided by Clarksons Research.
Notes: * Intermediate fuel oil with a maximum viscosity of 380 centistokes (<3.5 per cent sulphur).
All variations of MGO, MDO and IFO are traditional fuel types.
Alternative fuels encompass: LNG, LPG, methanol, biofuels, hydrogen, ammonia; synthetic methane and nuclear - highlighted in green.
Fuels that mention a traditional fuel type, along with an alternative fuel (for example: "Ethane, VLS IFO"; "Biofuel, VLS MGO" or "Nuclear, VLS MDO" refer to dual-fuel ships highlighted in light orange.

> **Box 2.1 Divided views on whether oil should be replaced by LNG**
>
> An alternative fuel already widely in use is liquefied natural gas. This is the greenest fossil energy source, which compared to heavy fuel oil (HFO), could reduce sulphur emissions by 99 per cent, nitrogen oxides by 80 per cent, and CO2 emissions by up to 20 per cent, along with most particulate matters. The 2020 Sphera report demonstrated that LNG/dual-fuel engines emit fewer grams of CO2 equivalent per kw than diesel engines. Dual-fuel engines can use existing technology, enabling ships to be operated on different types of fuel and comply with regulations while remaining competitive.
>
> In January 2021 the IMO sulphur cap entered into force, prompting greater investment in bunkering port infrastructure and in LNG-fuelled ships. Currently, these represent a small share of the fleet and of the orderbook. But their numbers are expected to grow significantly in the 2021–2022 period.
>
> The major disadvantage of LNG is that it consists primarily of methane which is a far more potent greenhouse gas than CO2. Even small escapes during production or use could result in a net increase in GHG emissions. In April 2021, the World Bank published a report that considered holistic lifecycle emissions and highlighted the impact of LNG on climate change. It recommended countries to avoid supporting LNG as a bunker fuel and advocated for regulation of methane emissions.
>
> Shipping industry voices, such as Maersk and Euronav, have also questioned the suitability of LNG as a transition fuel and point to the high costs of investing in new ships and infrastructure while not reducing lifecycle greenhouse gas emissions – with the danger of technological lock-in since new infrastructure with be in operation for 20 years. They also perceive such investment as extending the use of carbon in the maritime energy supply chain and delaying the energy transition.
>
> *Sources:* Gaztransport Technigaz (GTT). LNG as a marine fuel. Gilbert, P., Walsh C., Traut M., Kesieme U., Pazouki K. and Murphy A. (2018). Assessment of full life-cycle air emissions of alternative shipping fuels, *Journal of Cleaner Production*, Volume 172, 20 January 2018. Clayton, R. (2019). LNG will be transitional fuel for 2030, Nor-Shipping hears. Lloyds List News, 03 Jun 2019. Ovcina, J. (2020). Clarksons: 27 per cent of the order book to run on alternative fuels. *Offshore Energy*, 1/12/2020. Lloyd's Register (2021). The complexities of the fuel supply chain as we move towards zero-carbon. 20/01/2021. World Bank (2021). The role of LNG in the transition toward low-and zero carbon shipping. Lloyds List (2021). Is LNG really borderline greenwashing? *Lloyds' List Shipping Podcast*, 14/05/2021.

C. PORT SERVICES AND INFRASTRUCTURE SUPPLY

The past year has been very testing for port operations. The impacts of COVID-19, compounded by the Ever Given incident in the Suez Canal, have resulted in congestion and equipment shortages and have disrupted supply chains. Nevertheless, ports have remained operational and continued to serve diverse flows of trade. Their experience has confirmed the importance of preparing for the unexpected and of building resilience (box 2.2). But the COVID-19 crisis has also opened up new opportunities to diversify and to create better links between maritime and other modes of transport.

> **Box 2.2 Building port resilience UNCTAD experience**
>
> The UNCTAD TrainForTrade Port Management Programme helps ports in developing countries become more efficient and competitive. During the pandemic, the programme worked with other United Nations entities on a joint project to keep transport networks and borders operational – by implementing standards, guidelines, metrics, tools and methodologies to facilitate the flow of goods and services, while containing the spread of COVID-19. The project supports governments, including customs and other border agencies, port authorities, and the business community.
>
> This work includes a course on Building Port Resilience Against Pandemics which addresses four areas: crisis protocols and communications strategy; staff management, well-being, and resilience; technology preparedness; and cargo flow continuity.
>
> Discussions during the course indicate that building resilience requires significant changes in port operations. These would necessarily differ from country to country but this forum allows practitioners to discuss and exchange experience and ideas and explore responses and actions. They have concluded that port clients, operators and governmental entities can cooperate to improve their information systems – aiming for uniformity, consistency and predictability, while minimizing confusion and uncertainty at times of disruption.
>
> Key to the programme's success is South-South cooperation. Local instructors deliver training supported by experts from UNCTAD and other port partners.
>
> *Source:* Information provided by the UNCTAD TrainForTrade Port Management Programme.

2. Maritime transport and infrastructure

Ecommerce, smart logistic hubs and intermodal connections

During the pandemic, consumers sought a safe way to meet their needs, leading to a boom in online retail sales – which in 2020 amounted globally to $4.28 trillion. This trend is expected to continue: in 2022 e-retail revenues are projected to grow to $5.4 trillion (Statista, 2021).

These higher volumes, combined with expectations for rapid delivery, have boosted the demand for better logistic facilities – in particular for sufficient warehouses to store products along with space to fulfil and despatch orders, while also providing value-added services.

Indeed investment decisions and port planning are increasingly being influenced by the expectations of retailers and logistic operators – who are looking to reduce costs by using seaports close to warehousing or distribution facilities and their end markets (Drewry 2021). To avoid congestion and ensure rapid replenishment, ports can offer storage and warehousing capacity and space for modern logistics.

Ports are also investing in more technology for monitoring supply chains, detecting potential disruption and generally tracking shipments to their destinations. In 2021, several Asian ports, including Sichuan and Hainan in China, launched or announced investments in smart logistics (American Journal of Transportation, 2021 and South China Morning Post, 2021).

To maximize ecommerce logistics operations, port operators need to be able to handle data efficiently (Drewry, 2021). For this purpose, port logistic are increasingly relying on digitalization – for exchanging information among customers, partners, suppliers and other actors, and for offering new services (Logmore, 2019). For example, one of the world's largest global terminal operators, DP World, has acquired Syncreon, a global provider of supply chain services (van Marle, 2021).

To take advantage of ecommerce, ports also need to be well connected to their hinterlands. Using new technology they can become smart logistic hubs that connect maritime and other modes of transport – facilitating supply chain connections, domestically, regionally and internationally. These need to operate in a more agile, intermodal fashion at times of congestion and disruption (Schwerdtfeger, 2021). Box 2.3 describes how intermodal connections can be advanced by best practices, standards and regulations.

Box 2.3   Guidance and standards for intermodal operations

The UN Economic Commission for Europe (ECE) promotes best practices and standards for sustainable transport while also developing, and overseeing the implementation of, legal instruments. ECE aims to support inland freight transport, by improving traffic safety, environmental performance, energy efficiency, security and efficient service provision.

A recent ECE report, the *Handbook for National Masterplans for Freight Transport and Logistics*, provides guidance to governments on how transport and logistic services can, in post-pandemic times, contribute to economic development and recovery. The report highlights the critical importance for intermodal operations of intelligent transport systems (ITS) and telematics that enable operators to shift freight seamlessly across transport modes and networks – to plan routes and deliveries, and optimize cargo flows and the use of infrastructure.

Maximizing the benefits of ITS to transport operations will mean training the workforce for increased specialization and technological innovation, and supporting ITS research and development in cost-efficient solutions. At the same time, there needs to be significant investment, partly through public-private partnerships, in high-performance digital infrastructure, while ensuring efficient data exchange and interoperability.

It is also important to agree on legal instruments and standards. An example is the 1991 European Agreement on Important International Combined Transport Lines and related installations (AGTC). This agreement aims to make international combined transport in the ECE region more efficient and attractive to customers, by developing a common infrastructure quality standard for combined transport on the main European corridors. The framework's important nodal points include transport terminals, border crossing points, stations for exchanging wagon groups, gauge-interchange stations, and ferry links and ports. Facilitating modal shifts enables international freight movements while reducing the damaging environmental impacts from transporting international freight by road.

Implementing AGTC minimum standards is expected to strengthen critical Euro-Asian railway routes that can connect Central Asian landlocked ECE members to international markets. To avoid temporary closure of borders as a result of pandemics, ECE is also considering an agreement for uninterrupted operation of designated core lines of the network.

> **Box 2.3 Guidance and standards for intermodal operations** *(cont.)*
>
> At present, digital exchange between different modes of transport, sectors and countries is quite fragmented, so ECE is working on digital standards for harmonizing digital exchange of data and documents based on existing UN/CEFACT semantic standards and reference data models. These will allow for interoperability along multimodal supply chains, using a common foundation for converting data between modes of transport, sectors and authorities.
>
> Tests to prove the concept are taking place. For example, UN/CEFACT and FIATA experts have prepared a digital version of the FIATA multimodal Bill of Lading, aligned to the MMT RDM. Another test has focused on exports of wood and cellulose from Belarus to Central Europe via Ukraine, the Black Sea and the Danube, combining rail, road, river and maritime transport information exchanges. These tests demonstrate the benefits of seamless data exchange between different modal consignment notes and maritime bills of lading. Experts are also currently working on IMO/FAL forms in Ukraine with a view to using them along multimodal transport routes.
>
> *Source:* Inputs provided by the ECE Secretariat and ECE (2021) *Handbook for National Masterplans for Freight Transport and Logistics*.

Greener industrial port activities

The world is now embarking on the transition to greener energy. This will be costly. Halving shipping emissions by 2050 is estimated to require an annual average investment of between $40 to $60 billion between 2030 and 2050. Most of this is for producing alternative fuels such as ammonia, hydrogen, and methanol among others, while also developing new land-based infrastructure for storage and bunkering (Krantz et al, 2020).

The energy transition has major implications for ports. Less trade in oil will reduce revenue from storing and distributing fossil fuels. Preparing for a future without carbon fuels, ports are therefore aiming to develop new markets and value-added services (The Conversation, 2021 and Manners-Bell, 2021). And despite the pressures faced in 2020, many have maintained their plans for investing in environmental sustainability (IAPH-WPSP, 2021). These include production of alternative energy, infrastructure to import alternative fuels, and for bunkering and storage to facilitate onward distribution (table 2.12). Some ports have benefitted from infrastructure green recovery plans and others from incentives for foreign investment.

Table 2.12 Industrial port projects capitalizing on green opportunities to generate new revenue streams

Alternative energy	Bunkering infrastructure	Facilitating import of alternative energy and storage infrastructure
• Project to develop hydrogen-based exports from the Port of Fujairah (United Arab Emirates) • Project to develop offshore wind energy to generate hydrogen at North Sea Port (Belgium)	• Pilot hydrogen filling stations in the port of Antwerp • Proposed hydrogen infrastructure at Kobe, Chita, Yokkaichi and Hibikinada ports (Japan), capitalizing on existing hydrogen pipeline	• Project to develop a terminal in Germany for import and onward distribution of LNG, encompassing storage and ancillary services (Brunsbüttel Ports, Germany)

Sources: Argus Media (2021): Japan studies options to cut coastal shipping emissions. ArgusMedia, 2/7/2021. OffshoreWind.Biz (2021) Equinor, Ørsted, Boskalis Join AquaVentus Offshore Wind-to-Hydrogen Project 4/5/2021. Savvides; Nick (2021). Antwerp and CMB team up to launch multimodal hydrogen filling station. The Loadstar, 10/6/2021. Liebig; L. (2021). The United Arab Emirates is well placed to capitalise on the pivot to hydrogen 13/4/2021. Pekic, Sanja (2021). North Sea Port to get hydrogen pipeline network. *Offshore Energy*, 3/6/2021.

There is also now greater interest in smarter and greener ports. Beyond transforming ports into carbon-neutral ecosystems this means using new data environments and artificial intelligence to enhance competitiveness and sustainability. Some factors affecting the development of such ports are indicated in table 2.13.

Table 2.13 Factors affecting the development of smart green ports

Dimension	Influencing factors	Indicators of success
Greenness	Energy-saving and emission-reducing capability	Port's capability in saving energy and controlling pollutant discharges
	Pollution treatment capability	Responsiveness and degree in treating pollutants
	Efficient utilization of resources	Whether a port has the capability to utilize resources effectively to reduce resource waste
	Environmental protection concept and policy system	Knowledge and practices of port management personnel and policymakers in green concepts
Agility	Agile production capability	Port's capability in fully utilizing the limited resources and responding quickly to orders
	Comprehensive logistic capability. Levels of a port's comprehensive logistic services and supply	Whether a port adopts refined operation modes and has JIT capabilities
Personalization	Port-differentiated service levels	Levels of a port's services that are different from those at other ports
	Personalized service levels for customers	Levels of personalized services provided by the port to customers
	Emergency and quick response capabilities	Port's response capabilities to multiple emergencies and adjustability to changes
Cooperation	International port-shipping cooperation	Degree and model of international port-shipping cooperation
	Port-city integration	Port-city cooperation
	Cooperation between subsidiary and parent ports	Cooperation between subsidiary and parent ports (international dry ports, feeder ports and inland port areas)
Intelligence*	Intelligent production infrastructure and operation	Intelligence degree of port infrastructure operation and production
	Intelligent administration	Intelligence degree of port administration
	Intelligent facility security	Intelligence degree of port facility security
	Innovative R&D and technology application	Port's technical innovation R&D capability and degree of application
Liberalization	Liberalization of trade and economic policies	Port's liberalization degree in domestic and foreign trade
	Facilitation of logistics and customs clearance	Port's coordination with the Customs and quarantine departments and degree of cargo transportation facilitation
	Openness of investment and financing	Openness of a port in market investment and financing

Source: Chen, J.; Huang, Tiancun, Xie, X; Lee, P. and Hua, C. (2019). Constructing the Governance Framework of a Green and Smart Port. Journal of Marine Science and Engineering.

* Defined as "more modern intelligent technologies integrated into port working environments to improve port operations".

D. THE IMPACT OF COVID-19 ON PORTS: LESSONS FROM THE UNCTAD TRAINFORTRADE PORT MANAGEMENT PROGRAMME

The TrainForTrade Port Management Programme brings together a strong network of ports across several continents, for which the programme has continued to upgrade its Port Performance Scorecard (PPS). Each April member ports complete a survey on their performance in the previous calendar year. This provides valuable data for strategic planning within ports and for evidence-based policy analysis at regional and state levels.

The data are collected through 82 questions from which the PPS derives 26 agreed indicators under the following categories: finance, human resources, gender, vessel operations, cargo operations, and environment (table 2.14). The same approach has been used each year since the inception of the PPS in 2012 thus ensuring consistency and comparability over time.

For the current scorecard for the five-year period 2016–2020, 51 port entities provided 3,301 data points – an average of 98 data points per indicator. Around half of the ports were small, less than five million tons, or medium, between five million and 10 million tons. The annual volume throughput for the largest port in the sample was 80.9 million tons and for the smallest was 1.5 million tons. Two-thirds were landlord ports – owning the basic infrastructure and leasing it out to operators – or used a mixed model.

Table 2.14 Port Performance Scorecard indicators, 2016–2020

Category	Indicator number	Indicator	Number of values	Mean
Finance	1	EBITDA / revenue (operating margin)	98	33.1%
	2	Labour / revenue	102	22.9%
	3	Vessel dues / revenue	101	15.8%
	4	Cargo dues / revenue	101	36.7%
	5	Concession fees / revenue	91	13.7%
	6	Rents / Revenue	96	5.7%
Human resources	7	Tons / employee	108	65 054
	8	Revenue / employee	101	$189 180
	9	EBITDA / employee	97	$98 029
	10	Labour cost / employee	96	$32 985
	11	Training cost / wages	96	1.3%
Gender	12	Female participation rate – all categories	108	17.5%
	12.1	Female participation rate – management	108	42.0%
	12.2	Female participation rate – operations	100	16.0%
	12.3	Female participation rate – cargo handling	74	5.7%
	12.4	Female participation rate – other employees	46	29.1%
Vessel operations	13	Average waiting time (hours)	92	14
	14	Average gross tonnage per vessel	106	18 184
	15.1	Average of oil tanker arrivals	114	9.8%
	15.2	Average of bulk carrier arrivals	115	10.5%
	15.3	Average of container ship arrivals	114	30.7%
	15.4	Average of cruise ship arrivals	113	1.1%
	15.5	Average of general cargo ship arrivals	116	27.4%
	15.6	Average of other ship arrivals	114	22.5%
Cargo operations	16	Average tonnage per arrival (all)	117	8 162 t
	17	Tons per working hour, dry or solid bulk	77	317 t
	18	Tons per hour, liquid bulk	55	367 t
	19	Boxes per ship hour at berth	70	27
	20	Twenty-foot equivalent unit dwell time (days)	63	6
	21	Tons per hectare (all)	107	141 704 t
	22	Tons per berth meter (all)	113	6 482 t
	23	Total passengers on ferries	89	959 899
	24	Total passengers on cruise ships	92	91 068
Environment	25	Investment in environmental projects / Total CAPEX	54	6.3%
	26	Environmental expenditure/revenue	77	1.8%

Source: UNCTAD, based on data provided by selected member ports of the TrainForTrade network.
Abbreviations: CAPEX, capital expenditure; EBITDA, earnings before interest, taxes, depreciation and amortization.

2. Maritime transport and infrastructure

1. Impact of COVID-19 pandemic across the TFT port network

In 2020, the COVID-19 pandemic had a significant impact on ports worldwide. As well as creating health risks for port workers and seafarers in all regions it also substantially reduced the volume of trade. Between 2016 and 2018 cargoes had been growing at a median value of five per cent per year and revenues by six per cent. In 2020, however, volumes fell by 4 per cent and revenues by 9 per cent (figure 2.11). The impacts on individual ports are illustrated in box 2.4, by the experience of the Port of Gijon in Spain, and in box 2.5, by the port system in Peru.

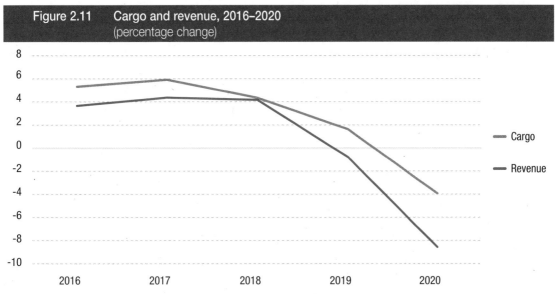

Figure 2.11 Cargo and revenue, 2016–2020 (percentage change)

Source: UNCTAD, based on data provided by selected member ports of the TrainForTrade network.

Box 2.4   Port performance analysis of the Port of Gijon in 2020

Although 2020 was a tough year for ports in general, and for Europeans in particular, for the port of Gijon it was what we could call the 'perfect storm'.

On the one hand, COVID-19 hit. On the other hand, the fight for a more sustainable world caused the closure of the five thermal power plants that the port served; consequently causing a loss of five million tons of coal. In addition, the shutdown of an Arcelor Mittal blast furnace caused a loss of almost four million tons.

Other traffic, such as the import, mix and export of coals from Russia to the Maghreb helped offset the large losses mentioned above. And despite the 'three storms', the Port of Gijon has firmly held the wheel while at the same time helping its clients, allowing them to delay payments for a year and rewarding companies affected by COVID-19.

Losses meant a 7 per cent drop over the previous year (2019). The total tons handled, amounted to 16 million tons. Traffic was broken down into 80 per cent solid bulks, 12 per cent general merchandise and 8 per cent made up of liquid bulks.

Iron ore, steel coal and cement made the port the first in solid bulk in the Spanish port system. Other solid bulks, like cereals and fertilizers, contributed to its leadership.

As for general merchandise, 75 per cent was containerized, with 85,000 TEU moved. This represented 75 per cent of the port's hinterland and is expected to expand in the coming years following a new rail connection with the centre of the country. The remaining 25 per cent of the total, 1.5 million ton of general merchandise, was steel products.

Liquid bulks represented 8 per cent of the mix – petroleum products, gasoline, and gasoil, intended for final consumption.

Despite the wind and seas from the bow, financial results have been positive and increased by a little over two million euros. The year 2021 is born full of new projects and hopes that will undoubtedly help turn the page of these challenging times.

Source: Port Authority of Gijon.

> **Box 2.5   Port performance analysis of the national port system in Peru in 2020**
>
> In Peru, in 2020 there was a 10.9 per cent fall in volumes to 97.4 million tons while the number of containers (in TEU) handled remained stable nationally. However, there was a drop in container traffic at the larger international terminals of 3 per cent compared to 2019 due to the impact of the COVID-19 health emergency (see table).
>
> The main types of goods, containers, solid bulk, and break-bulk cargo, decreased by 0.3 per cent, 3.7 per cent and 4.9 per cent, respectively, as shown in the table, which illustrates the movement of cargo at public and private port terminals for 2019/2020.
>
> | Type of Merchandise | Unit of measure | Year 2019 | Year 2020 | Change (%) 2020/2019 |
> |---|---|---|---|---|
> | | TEU | 2 678 258 | 2 654 289 | -0.9% |
> | LoLo containers | units | 1 618 433 | 1 592 256 | -1.6% |
> | | tons | 25 905 625 | 25 832 736 | -0.3% |
> | Break Bulk | tons | 4 057 174 | 3 858 419 | -4.9% |
> | Bulk Solids | tons | 12 165 301 | 11 714 440 | -3.7% |
> | Bulk Solid Minerals | tons | 33 122 675 | 27 978 125 | -15.5% |
> | Liquid Bulks | tons | 33 756 658 | 27 883 897 | -17.4% |
> | RoRo | tons | 333 213 | 207 063 | -37.9% |
> | **Total Load** | tons | 109 340 647 | 97 474 680 | -10.9% |
>
> However, these reductions are moderate compared to those for bulk minerals, liquid, and roro cargo, which decreased by 15.5 per cent, 17.4 per cent, and 37.9 per cent, respectively.
>
> During the year 2020, the National port system handled a total of 2.6 million TEUs, presenting a slight drop of 0.9 per cent, compared to the year 2019.
>
> *Source:* National Port Authority of Peru.

Financial performance

Financial performance of ports can be measured as the average gross revenue per ton of cargo. This ranged from $1.9 per ton in Europe, and $2.26 in Asia, to $5.31 in Africa. At the global level the sources of revenues are indicated in figure 2.12, showing the split between port dues on vessels and cargo throughput, port service charges, and income derived from land and concession rights.

Around half of revenues come from vessel and cargo charges for the use of primary port infrastructure. This proportion is likely to fall over time with the development of digitalized ports and energy hubs, using either the concession or landlord model.

Profitability is measured as earnings before interest, taxes, depreciation, and amortization (EBITDA). Businesses with high demands for infrastructure investment require elevated levels of EBITDA to be sustainable. In 2020 average profitability declined by 12 per cent in Europe, by 17 per cent in Asia and by 25 per cent in Africa. Latin America showed no change. These declines can be partly explained by the impacts of COVID-19, though in Africa there must be other major factors since volumes and revenues showed only a minor impact from the pandemic.

In performance terms, the reported numbers show a falloff in 2020. While there have been profitability drops in other periods this decline can be partially explained by the COVID-19 pandemic.

Last year the scorecard covered the period 2015–2019, for which EBITDA as a proportion of revenue was 38.8 per cent (indicator 1). The 2021 scorecard covered the period 2016–2020 for which the proportion

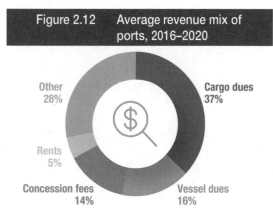

Figure 2.12   Average revenue mix of ports, 2016–2020

- Cargo dues 37%
- Other 28%
- Vessel dues 16%
- Concession fees 14%
- Rents 5%

Source: UNCTAD, based on data provided by selected member ports of the TrainForTrade network.

declined to 33 per cent. The impact was, however, lower in Europe where averages remained at 59 per cent and in Latin America at 41 per cent.

A high-level comparison of revenue profiles shows the mix between port dues on vessels and cargos, port service charges and incomes derived from lands and concession rights. Between 2020 and 2021 scorecards, the proportion of total capital expenditure for environmental purposes fell from 7.2 to 6.4 per cent, while the proportion of operating costs for environmental purposes fell from 2.3 to 1.8 per cent. In some countries the environmental data are difficult to extract since they can be embedded in the total capital or operating spends.

Gender equality

Sustainable Development Goal 5.5 calls for full and effective participation of women and equal opportunities for leadership at all levels of decision making in political, economic, and public life. In this respect, ports still do not perform well. Between 2020 and 2021 scorecards, the average female proportion of the port entity workforce fell slightly, from 17.6 to 17.5 per cent. The proportion in Europe is significantly higher at 24.8 per cent, though most of these women work in management or administration.

Overall, the figures are more encouraging for management and administrative roles. Between 2020 and 2021 scorecards, the proportion of women rose from 38 to 42 per cent. Asia led the way at 52 per cent, followed by Europe at 39 per cent. Female participation is however far lower for cargo handling and port operations. There is thus still a lot to be done to achieve the SDG target to "Achieve gender equality and empower all women and girls." Box 2.6 illustrates how the Philippines Ports Authority is making the changes to meet this objective.

Box 2.6   Gender and development in the Philippine Ports Authority and its journey

The Philippine Ports Authority (PPA), under the present leadership of Atty. Jay Daniel R. Santiago, General Manager, has continued its commitment to institutionalize gender and development (GAD) in all the ports under its jurisdiction. For SDG 5: "Gender Equality" PPA now satisfies target 5.5 "Ensure women's participation and leadership in decision-making".

The port industry in the Philippines is undeniably male-dominated. However, in recent years, women have been making remarkable progress, within the Authority particularly at management levels and PPA continues to put a premium on women's empowerment. In its GAD journey there have been many firsts in entrusting some of highest managerial positions to female officers: first female Assistant General Manager on Finance and Administration (executive level); first female Port Manager (managerial level in field offices); and first female Department Manager (managerial level in head office).

As of May 2021, women made up half of PPA's workforce, amounting to 1,026 female personnel. The highest women-occupied positions are at the middle management level with two department managers, five port managers and 56 division managers. Some women employees are also taking male-dominated positions such as terminal supervisor, safety officer, civil security officer, engineer, terminal operations officer, or industrial security officer. This shows that the authority values the immense contributions of women employees in the areas of decision-making, management, operations, and even security.

To further strengthen GAD initiatives the Authority ensures compliance with statutory laws upholding the welfare and development of Filipino women. For instance, PPA strictly observes the provisions of the General Appropriations Act and Republic Act 9710, also known as the Magna Carta of Women, which directs government agencies to formulate a GAD plan, the cost of which shall be not less than 5 per cent of the annual budget. Annually, PPA appropriates 5 per cent of its corporate budget for implementing the Authority's GAD plans and programmes. Among the GAD flagship projects and programmes are the construction of gender-neutral facilities and halfway houses, along with capacity-building to increase awareness among employees.

In recent years, PPA has been crafting and implementing gender-responsive policies, plans and programmes to advocate gender equality and women's empowerment. This has been given an added impetus by the UNCTAD TrainForTrade port management programme in the Philippines. Many women have participated in the three cohorts of the programme and more are expected to join subsequent cycles.

Source: Philippine Ports Authority.

Vessel and cargo operations

PPS data provide interesting insights into the differences between regions. At the global level, for the 2016–2020 period, compared with the previous five-year average, the average cargo load per vessel per arrival rose from 7,865 to 8,162 tons, a 3.9 per cent increase (indicator 16). However, these average loads vary greatly between regions, reflecting different types of operations and distances to market. Asia, for example, has a high proportion of passenger ferry operations and an average load of only 2,313 tons, while Africa on average has longer journeys made by larger vessels and an average load of 15,681 tons. Globally, there was little change in average vessel size which rose from 18,124 to 18,184 Gross Tons (GT) (indicator 14) in the 2016–2020 period compared with the previous five-year average (2015–2019).

One of the most direct impacts of COVID-19 was on the number of passengers. For the 2016–2020 period, compared with the previous five-year average, passenger numbers fell by 34 per cent (indicator 23). There was similar fall in the number of cruise passengers, by 28 per cent (indicator 24). Between years 2019 and 2020 only, the number of passengers on ferries fell by 71 per cent and on cruise ships by 76 per cent.

Overall, modern ports show many similarities in their financial and operations data as well as in their declared policy profiles and corporate structures. Nonetheless, each port entity has its own unique characteristics. Some may have greater autonomy on pricing while for others this might require national-level approval. Control over major investments, however, appears to be retained at the political level.

The pandemic has accelerated digitalization and decarbonization and key theme of future data analysis will be on how performance levels are affected by such changes.

E. SUMMARY AND POLICY CONSIDERATIONS

This chapter has provided recent information in some key areas:

- *Fleet size* – Between 1 January 2020 and 1 January 2021, the world fleet grew at the historically low rate of 3 per cent, reaching 99,800 ships of 100 gross tons and above, equivalent to 2.13 billion dwt of capacity in January 2021. Ships delivered in 2020 were mostly bulk carriers, followed by oil tankers and container ships. During this period, ship deliveries declined by 12 per cent, partly due to lockdown-induced labour shortages for marine-industrial activity. The number of ships sold for recycling increased in 2020, although levels remained low by historical standards.

- *Ship orders* – During 2020, ship ordering declined by 16 per cent, continuing the downward trend observed in previous years, though newbuilding orders surged during the first half of 2021. As owners and operators tried to cope with tight vessel supply, they turned to the second-hand market, leading to higher second-hand prices. In several shipping segments, the current imbalance between supply and demand has pushed up freight rates.

- *Regulation* – Regulatory changes to align shipping operations with decarbonization targets, along with the energy transition creates an uncertain environment that will affect shipping, trade and energy use and entail significant costs. The short-term measure agreed recently at the IMO could affect ship costs, ship travel distance, fleet distribution, routing patterns, and use of different types of vessels and may increase maritime logistics costs. Slow steaming to reduce fuel consumption could result in the need to increase the number of ships.

To cater for the high demand for ships, shipping companies will need to expand their fleets and scale up investment. Meeting the decarbonization target will require retrofitting or replacement. In developing countries in particular it will be important to assess the implications of regulatory measures. For replacing older vessels with larger and more fuel-efficient ships and making the corresponding landside investments, investors will need more predictable regulatory environments, and greater certainty when trialling and scaling up alternative fuels.

While adding to the pressures, the pandemic has often accelerated necessary changes. Many ports for example, are embracing new strategies, capitalizing on ecommerce opportunities and preparing for a future without carbon fuels by embarking on greener industrial port activities – evolving into green smart ports that can become catalytic hubs for revenue generation and industrial growth. Key to all these changes is digitalization which is redefining port business success and facilitating intermodal operations. Both seaports and inland ports will need support to keep up with digitalization, so as to function efficiently and seize opportunities as they arise.

REFERENCES

American Journal of Transportation (2021) Kalmar enters into strategic cooperation with Sichuan Port and Shipping Investment Group in China. 09 June.

Allianz Global Corporate and Speciality (2019). Loss trends – larger vessels bring bigger losses. Expert Risk Articles, News and Insights.

Bak M. Newbuild Investment: On The Up Again.., Clarksons *Shipping Intelligence Network*. 23 July 2021.

Balcombe P, Brierley J, Lewis C, Skatvedt L, Speirs J, Hawkes A and Staffell I (2019). How to decarbonise international shipping: options for fuels, technologies and policies. Energy Conversion and Management Journal, 182:72–88.

Boulougouris E (2021). Suez Canal blockage: how cargo ships like Ever Given became so huge, and why they are causing problems. The Conversation, 01 April.

Chen J, Huang T, Xie X, Lee P and Hua C (2019). Constructing the Governance Framework of a Green and Smart Port. Journal of Marine Science and Engineering.

Clarksons Research (2021a). *Container Intelligence Monthly*, Volume 23, No. 6. June.

Clarksons Research (2021b). Shipping Intelligence Weekly Issue No. 1,478, 25 June.

Clarksons Research (2021c). World Shipyard Monitor. Volume 28, No. 1 January.

Clarksons Research (2021). Tracking "Green" Technology Uptake. June 2021.

Clayton R (2019). LNG will be transitional fuel for 2030, Nor-Shipping hears. Lloyds List News, 3 June.

Drewry (2021). Webinar - Container Ports & Terminals, June 2021.

ECE (2021). *Handbook for National Masterplans for Freight Transport and Logistics*.

Gaztransport Technigaz (GTT). LNG as a marine fuel.

Gilbert P, Walsh C, Traut M, Kesieme U, Pazouki K and Murphy A (2018). Assessment of full life-cycle air emissions of alternative shipping fuels, *Journal of Cleaner Production*, Volume 172, 20 January 2018.

Hayden R (2015). Mega-ships accidents pose risks to supply chains, shippers, insurers. JOC, 30 September.

Haralambides H, Tsolakis S, and Cridland C (2005). Econometric Modelling of Newbuildings and Secondhand Ship Prices.

Hellenic Shipping News (2021). The Suez Canal Accident and the State of Global Shipping, Port News, 19 June.

ING (2021). European power markets ease but prices remain high. 27 July 2021.

Jiang J (2021). IMO 2020 and demolition volumes Splash 24/ 7.

IAPH-WPSP (2021). Port Economic Impact Barometer One Year Report. May 2021.

IMO (2018). Note by the IMO to the UNFCCC Talanoa Dialogue: "Adoption of the initial IMO strategy on reduction of greenhouse gas emissions from ships and existing IMO activity related to reducing greenhouse gas emissions in the shipping sector".

Kachi A, Mooldijk S and Warnecke C (2019). Carbon pricing options for international maritime emissions.

Krantz R, Sogaard Kand Smith T (2020). The scale of investment needed to decarbonize International Shipping Getting to Zero Coalition, January 2020.

Lockton (2019). Re-evaluating the risk of mega ships. Lockton Insights, UK.

Logmore (2019). Digital Logistics: What Is It, and How Will It Impact Your Organization? 16 April.

Lloyds List (2021). Is LNG really borderline greenwashing? *Lloyds' List Shipping Podcast*. 14 May.

Lloyd's Register (2021). The complexities of the fuel supply chain as we move towards zero-carbon. 20 January.

Manners-Bell J (2021). Ports plan transition to hydrogen superhighway hubs. Transport Intelligence. 17 June.

Maritime Executive (2021). Global Orderbook at Seven-Year High After Strong Start to 2021. 6 July 2021.

Market and Research News, 2021. China Shipbuilding Industry Report 2021–2025: Depending on COVID-19 Impacts in 2021, the Industry May Not Recover Until 2022. 15 February 2021.

Medeiros E, Cliff G, Crane K and Mulvenon, J (2021). A New Direction for China's Defense Industry, Chapter: China's Shipbuilding Industry. Rand Corporation.

Miller G (2021). No letup yet in dry bulk shipping's 'remarkable rally'. Freight waves. 28 June.

OECD (2019). Ship recycling: an overview. OECD Directorate for Science, Technology and Innovation.

Organization of Economic Cooperation and Development and International Transport Forum (2018). Decarbonizing Maritime Transport: Pathways to Zero-carbon shipping by 2035.

Ovcina J (2020). Clarksons: 27 per cent of the order book to run on alternative fuels. *Offshore Energy*. 1 December.

Ovcina J (2021) Stopford: Industry will need $3.4 trillion in the next 30 years to replace existing fleet. *Offshore Energy*. 15 April.

Plevrakis G (2020). Decarbonisation: energy market scenarios and the impact on the shipping fleet. Safety4Sea. 20 May.

Roussanoglou N (2021a). Ship Acquisitions in the Dry Bulk Market Are Booming. Dry Bulk Market News in Hellenic Shipping News. 15 June.

Roussanoglou, N (2021b). Ship Owners Order more Container Ships and LNG Carriers. Hellenic Shipping News Worldwide. 29 July 2021.

Sancricca M (2016). Second hand ships vs. new buildings: a financial perspective. 20 November.

Sanchez R (2021). Looking into the future ten years later: big full containerships and their arrival to South American ports. Journal of Shipping and Trade 6, 2.

Sanchez R, Sanchez S and Barleta E (2020). Towards the decarbonization of international maritime transport: findings from a methodology developed by ECLAC on shipping carbon dioxide emissions in Latin America. UNCTAD Transport and Trade Facilitation Newsletter N°86 - Second Quarter 2020, Article No. 56. 17 July.

Schwerdtfeger M (2021). Intermodal traffic will shape the future of US container ports. Port Technology. 21 April.

Shell; Deloitte. Decarbonising Shipping: All Hands-on Deck; Shell: The Hague, The Netherlands, 2020.

ShipInsight (2021). Container ship ordering continues record upward trend. 22 July 2021.

Springer A (2021). Newbuild Prices: On The Up In 2021 So Far. Clarksons *Shipping Intelligence Network*. 23 June.

Statista (2021). Retail e-commerce sales worldwide from 2014 to 2024.

The Conversation (2020). How shipping ports are being reinvented for the green energy transition. 20 June.

The Maritime Executive (2021). Ordering Surges at Asia's Biggest Shipbuilding Yards. 30 May.

UNCTAD (2021a). Container shipping in times of COVID-19: why freight rates have surged, and implications for policymakers.

UNCTAD (2021b). Assessment of the impact of the IMO Short-Term greenhouse gas reduction Measure on States: Assessment of impacts on maritime logistics cost, trade and GDP.

van Marle G (2021). DP World snaps up contract logistics operator Syncreon for $1.2bn. The Loadstar. 1 July.

Walia I (2021). Dry bulk higher earnings are 'a predawn', says Lindström. Lloyds' List News. 29 June.

World Bank (2021). The role of LNG in the transition toward low-and zero carbon shipping.

Zhang J (2021). Alibaba logistics arm Cainiao partners with Hainan to bring smart logistics to duty-free island province. South China Morning Post. 10 June.

This chapter reports on recent developments in freight rates and transport costs. It covers 2020 and the first half of 2021, tracking changes in demand and supply across key shipping markets. It considers the immediate outlook for freight markets and examines the impact on prices.

As indicated in previous chapters, the COVID-19 pandemic led to a sudden dip in international seaborne trade. But by late 2020 there had been a swift rebound mainly in container and dry bulk shipping. The recovery in container trade flows, which was mainly on East-West containerized trade lanes, created a series of logistical challenges and hurdles, pushed up rates and prices, increased delays and dwell times, and undermined service reliability. As a result, there have been calls for more government intervention and regulatory oversight to mitigate any unfair market practices.

Sustained higher container freight rates would increase costs in global supply chains which could work their way through to higher consumer prices, with adverse economic effects globally – but particularly on the small island developing states (SIDS) and the least developed countries (LDCs) whose consumption and production depend more on international trade. There have been similar surges in trade and prices for dry bulk freight. The situation for tanker shipping, however, has been very different: a drop in global fuel demand and high carrying capacity have pushed tanker rates to record lows.

This chapter also highlights the structural determinants that shape transport cost such as port infrastructure, trade facilitation measures, liner shipping connectivity, and bilateral trade imbalances.

3

Freight rates, maritime transport costs and their impact on prices

Freight rates, maritime transport costs and their impact on prices

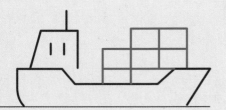

MARITIME FREIGHT RATE MARKETS

Record-breaking freight rate levels

As of late 2020 and into 2021 freight rates surged across **containerised and dry bulk shipping markets** and hit record highs

Tanker markets came under pressure with tanker rates reaching low levels

Container freight rates
Skyrocketed amid surge in demand for container shipping and limited capacity including container shortages and congestion at ports ↑

Dry bulk freight rates
Reached record breaking levels, driven by solid growth in demand that exceeded fleet growth ↑

Tanker freight rates
Fell to record lows as global fuel demand decreased and the supply of vessel carrying capacity remained high ↓

SIMULATED IMPACT OF CONTAINER FREIGHT RATE SURGES
Hardest hit will be SIDS

Simulation assumption:
Sustained increase in container freight rates

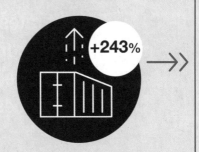 **+243%** →

Simulation results:

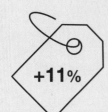

Increase in global import price levels
+11%

Increase in consumer price levels by country groupings

LLDC	World	LDC	SIDS
+0.6%	+1.5%	+2.2%	+7.5%

SIMULATED IMPACT OF IMPROVING MARITIME TRANSPORT COST DETERMINANTS

Simulation is conducted using the new dataset developed by UNCTAD and the World Bank

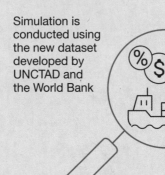

Simulation assumption:
Improving structural determinants ↓

Simulation results:
Reduction in maritime import transport costs

Port infrastructure	Trade facilitating environment	Shipping connectivity
-4.1%	-3.7%	-4.4%

A. RECORD-BREAKING CONTAINER FREIGHT RATES

In 2020, lockdown measures and other impacts of COVID-19 suddenly cut the demand for containerized goods. April and May 2020 were the worst months: by the end of May 2020, a record 12 per cent of global container capacity was idle or inactive – 2.7 million TEU (BIMCO, 2020. Clarksons Research, 2021a). Liner shipping companies responded with measures to mitigate costs, manage capacity and sustain freight rates. By the second half of 2020, the situation had reversed, but this sudden boost in demand stumbled into limited capacity and congested ports.

1. In mid-2020 high demand and limited capacity led to rocketing spot freight rates

In the second half of 2020, demand for container shipping started to pick up and absorb spare capacity. Vessel supply capacity remained limited but idle container shipping capacity levels started to decline in line with growing demand as trade continued to recover. By the end of June 2020, idling was 9 per cent, but by July this proportion had fallen to 6 per cent, and by August to 4 per cent. By the end of September 2020, it was down to 3.5 per cent (going below the 4.1 per cent average level of idling for full year 2019) (Clarksons Research, 2021a).

In 2020, global container fleet capacity expanded by almost 3 per cent, to 281,784,000 dwt (see also chapter 2), while container trade contracted by 1.1 per cent to 149 million TEU (figure 3.1).

Figure 3.1 Growth of demand and supply in container shipping, 2007–2021, percentage

Source: UNCTAD secretariat calculations. Demand is based on data from chapter 1 – figure 1.5, and supply is based on data from Clarksons Research, *Container Intelligence Monthly*, various issues.
Notes: Supply data refer to total capacity of the container-carrying fleet, including multipurpose and other vessels with some container-carrying capacity. Demand growth is based on million TEU lifts.

In an effort to maintain freight rates during the period of lower demand, carriers restricted capacity. Then as demand picked up, they released more capacity but by that time the supply was being constrained by other factors, notably port congestion and equipment shortages which kept vessels waiting, especially in West Coast North America. The result was exacerbated disruption and inefficiency at port.

By the end of 2020, freight rates had surged to unexpected levels. This was reflected in the China Containerized Freight Index (CCFI) for both short- and long-term contracts (figure 3.2). In the second quarter of 2020, the CCFI stood at 854 points, but by the fourth quarter was 1,250 points, and for the first and the second quarters of 2021 had reached new records, beyond 2,000 points.

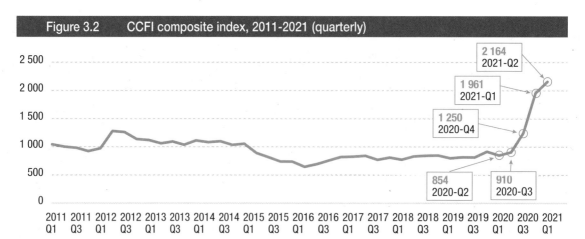

Figure 3.2 CCFI composite index, 2011-2021 (quarterly)

Source: Clarkson *Shipping Intelligence Network* Timeseries, Shanghai Shipping Exchange.
Note: The CCFI tracks spot and contractual freight rates from Chinese container ports for 12 shipping routes across the globe, based on data from 22 international carriers.

2. Container shortages, port congestion and delays result in higher freight rates, fees and surcharges

Towards the end of 2020 and into 2021, container shortages and congestion at ports, along with other disruption, led to record container freight rates, notably on the routes from China to Europe and the United States. These are reflected in the Shanghai Containerized Freight Index (SCFI) which covers cargo departing from Shanghai, China (figure 3.3). In June 2020, SCFI spot rate on the Shanghai-Europe route was less than $1,000/TEU but by the end of 2020 had reached around $4,000/TEU and remained firm throughout the first quarter of 2021. By the end of April, despite a 3 per cent increase in supply capacity (Clarksons Research, 2021a), the SCFI spot freight rate on the Shanghai-Europe route surged to $4,630/TEU, and by the end of July has reached $7,395/TEU.

Freight rates also escalated on the China-United States trade lane, and, faced with backlogs and longer waiting times, shipping lines have also been adding extra fees and surcharges. In the last quarter of 2020, on the Shanghai-West Coast North America route capacity expanded by 5 per cent and in the first quarter of 2021 by a further 7 per cent (Clarksons Research, 2021a). Nevertheless, the SCFI spot rate reached around $4,500/ forty-foot equivalent unit (FEU) in April 2021, compared to $1,600/FEU in April 2020, and climbed further to $5,200/FEU in July 2021. The trend was similar on routes from Asia to the East Coast. In the first six months of 2021, SCFI spot rates on the Shanghai-East Coast North America route more than doubled, and

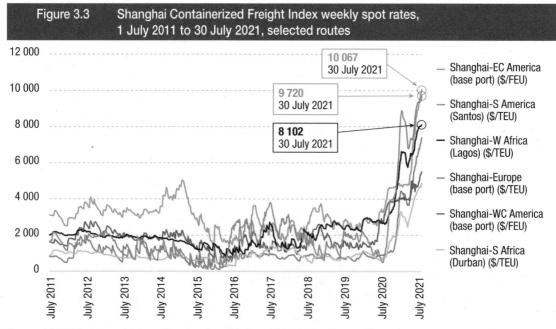

Figure 3.3 Shanghai Containerized Freight Index weekly spot rates, 1 July 2011 to 30 July 2021, selected routes

Source: UNCTAD secretariat, based on data from Clarkson *Shipping Intelligence Network*.

by the end of July 2021 had reached $10,067/FEU (figure 3.3). Moreover, this does not take into account the premiums cargo owners were often charged to get any certainty that their boxes would be moved promptly.

The surge in spot freight rates also extended across developing regions, including South America and Africa. On the China to South America (Santos) route the rate had been $959/TEU in July 2020 but by the end of July 2021 had reached $9,720/TEU. Over the same period, rates on the Shanghai to West Africa (Lagos) route increased from $2,672/TEU to $8,102/TEU. There was also a surge in rates from China to the Arab region. Box 3.1 provides further information on the impact of COVID-19 on maritime freight in the Arab region.

Box 3.1 Impact of COVID-19 on maritime freight rates in the Arab region

Fluctuations in freight rates reflect changes in lockdown policies and varying speeds of recovery, as well as the impact of shortages of both containers and ships and congestion in key ports and shipping nodes. These surges are likely to be amplified in most of the low- and middle-income countries of the Arab region, especially those suffering from conflicts or economic or financial crises which have had major impacts on patterns of production and consumption – and on maritime freight rates. Between October 2020 and June 2021 the SCFI from Shanghai to Dubai rose by 176 per cent and from Shanghai to the Mediterranean ports by 400 per cent.

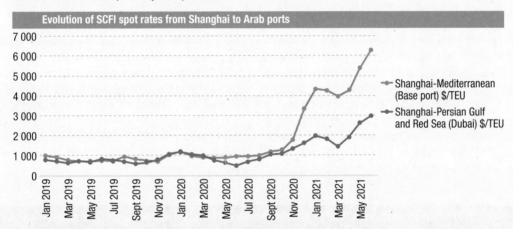

Source: UNCTAD/ESCWA calculations based on data from Clarksons Research.

To alleviate the impact on consumer prices, some countries have adopted special measures. In Lebanon, for example, when calculating the customs fees on imported goods, the customs authorities are still using the official exchange rate, which is far below the black-market exchange rate. In Jordan, when calculating customs fees on imported goods, customs authorities have put a ceiling on freight rates. According to the International Chamber of Navigation in Beirut, both measures did slightly alleviate the impact on consumers. But these subsidies may be difficult to sustain, so it will be important to consider the economic and financial evidence, to see how they compare with more conventional trade facilitation procedures.

There have also been initiatives to address the impact of COVID-19 at the regional level. In October 2020, ESCWA/UNCTAD published a working paper 'COVID-19: Impact on Transport in the Arab Region', which was summarized in a policy brief. On 24 November and 8 December there was a remote round table within the activities of the 21st session of ESCWA committee on Transport and Logistics. This was serviced by a parliamentary paper on the 'Impact of the COVID-19 pandemic on transport in the Arab region'.

In addition, in partnership with UNCTAD and other UN regional commissions, ESCWA implemented several activities within the UN Development Account project on transport and trade connectivity in the age of pandemics. This included producing material on 'Coronavirus Disease (COVID-19): Trade and Trade Facilitation Responses in the Arab Region' as well as a report on the 'Collective Application of eTIR Across a Land Transport Corridor Connecting East Mediterranean to GCC countries (Lebanon-UAE)'. On 16–17 December 2020, in cooperation with ECE, International Road Transport Union (IRU) and the Euromed Transport Support Project, ESCWA developed three questionnaires for banks, firms and policy makers aimed at gauging the conditions for trade financing in the region.

ESCWA also organized an online capacity building workshop on 'Implementation of the eTIR International System in the ESCWA region'. Also, in cooperation with ECE, it helped connect the national customs system of Tunisia to the international eTIR system.

Finally, ESCWA has provided substantive support and input to the initiative led by the Department of Transport and Tourism of the League of Arab States on addressing the impact of COVID-19 – with recommendations that were categorized according as short term (containing), medium term (recovery) and long term (resilience to future crises). These recommendations were adopted by the 33rd session of the Council of the Arab Ministers of Transport, held in Alexandria, Egypt, on 21–22 October 2020.

Contribution from ESCWA.

High shipping costs arising from logistical bottlenecks and lack of containers and equipment

Since late 2020, shipping costs have increased in part because of a shortage of containers. Containers are shipped full from export-oriented locations, notably in Asia, and many usually return empty. As Asia slowly began to recover, other countries remained under national lockdown and restriction so the importing countries could not return containers. The resulting shortage of empty containers was exacerbated as carriers introduced blank sailings where empty containers were left behind and failed to be repositioned. These impediments led to higher container dwell times at ports, and empty containers not returning to the system where they were most needed (UNCTAD, 2021). This increased shipping costs as shippers were reported to be paying premium rates to get containers back (CNBC, 2021), in addition to surcharges arising from port congestion and delays, including delays in returning equipment.

With containers scarce and ports suffering from congestion, shippers, freight forwarders, and importers were charged increasingly higher demurrage and detention fees. Between 2020 and 2021, across the world's 20 biggest ports, the average demurrage and detention charge doubled – equivalent to $666 for each container (Container xChange, 2021).

3. Surge in spot freight rates leading to increases in contracted rates

An important part of containerized trade is carried out at confidential contract rates negotiated between shippers and shipping lines. These rates are influenced by prevailing market conditions so in 2021 when spot rates were high, contract rates were correspondingly high and some were negotiated quickly to secure deals. Shipping lines typically gave priority to larger and more established shippers – leaving out smaller ones who were often unable to renegotiate. For their part, shippers aiming to hedge against future increases and uncertainties were increasingly seeking multi-year contracts. In 2021, many shippers signed trans-Pacific volume contracts for between $2,000/FEU and $3,000/FEU (Hellenic Shipping News, 2021b) – far higher than previous rates on the same routes. See also table 3.1 on contract freight rates which includes all surcharges including terminal handling charges.

Table 3.1 Contract freight rates, inter-regional, 2018–2020, $ per 40-foot container (FEU)

From	To	Average	2018	2019	2020
Africa	Africa	1 862	1 812	1 849	1 924
	Asia	758	748	750	775
	Europe	1 607	1 431	1 643	1 747
	Latin America	1 950	2 010	1 860	1 979
Asia	Africa	1 946	1 800	1 927	2 112
	Asia	768	737	747	821
	Europe	1 848	1 782	1 847	1 916
	Latin America	2 198	2 290	2 075	2 230
	North America	2 580	2 426	2 603	2 711
	Oceania	1 803	1 770	1 790	1 850
Europe	Africa	1 701	1 595	1 650	1 858
	Asia	947	967	870	1 004
	Europe	887	804	881	976
	Latin America	1 232	1 019	1 302	1 376
	North America	1 838	1 518	1 742	2 256
	Oceania	2 002	1 996	1 933	2 077
Latin America	Africa	1 910	1 778	1 951	2 000
	Asia	1 796	1 623	1 963	1 802
	Europe	1 751	1 313	1 977	1 961
	Latin America	1 529	1 349	1 699	1 539
	North America	1 716	1 521	1 882	1 745
North America	Africa	2 994	2 890	3 112	2 981
	Asia	1 129	1 009	1 111	1 269
	Europe	1 097	858	1 109	1 323
	Latin America	1 353	1 254	1 318	1 486
	North America	1 516	1 534	1 429	1 584
	Oceania	2 722	2 538	2 634	2 996

Source: UNCTAD, based on data provided by TIM Consult Market Intelligence https://timconsult.com/service_areas/transport/benchmarking/.

Note: The data set provides regional averages for forty-foot container dry cargo freight, as negotiated for routes where rates were available for at least 5 shippers and at least 500 TEU per year on port-pair basis.

Rates are "gate-in gate-out", i.e., including terminal handling charges and all charges and surcharges of ocean transport. Not included are pre- and on-carriage as much as classical administrative services of forwarders (customs clearance, booking and invoice control fees, etc.). The average is unweighted, based on representative main ports. Trade imbalance is also impacting freight rates.

The new data set, provided by TIM Consult Market Intelligence as per table 3.1, enables an overview of actual basic freight rates on different routes, including inter-regional routes, and their development over time.[1] Imbalanced trade flows mean that transport costs tend to be higher in the direction of the high-demand region thereby impacting freight rates (Jonkeren, Olaf, et al, 2011). Between 2018 and 2020, rates on the Asia-Europe leg, for example, were twice as high as those on the Europe-Asia leg. Similarly, rates for exports from Asia to North America were twice as import rates. As for the Asia-Africa trade the ratio was 2.6, and intra-African freight rates were 2.4 times higher than intra-Asian rates. Over this period the most volatile rates were those to and from Latin America.

4. Trends in charter market rates in sync with spot freight rates

In the first half of 2020, the COVID-19 crisis also reduced container ship charter rates, especially for larger ships. This was a period of falling demand, ship idling, capacity withdrawal, and blank sailings. But the situation reversed in the second half of 2020 with increasing demand for ships of all sizes. In June 2020, the New ConTex index fell to 308 points but by December 2020 had more than doubled to 687 points (figure 3.4). In 2021, the continuing imbalance between demand and supply pushed the ConTex average to unforeseen levels reaching 1,645 points in June and 2,348 in July.

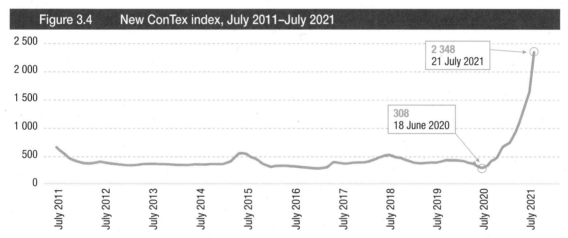

Figure 3.4 New ConTex index, July 2011–July 2021

Source: UNCTAD secretariat, based on data from the New ConTex index produced by the Hamburg Shipbrokers Association. See http://www.vhss.de (Accessed on 25 July 2021).

Notes: The New ConTex is based on assessments of the current day charter rates of six selected container ship types, which are representative of their size categories: Type 1,100 TEUs and Type 1,700 TEUs with a charter period of one year, and Types 2,500, 2,700, 3,500 and 4,250 TEUs with a charter period of two years.
Index base: October 2007 – 1,000 points.

5. Container shipping profits are high, as are short and medium terms freight rates

High freight rates have boosted the profits of global container shipping companies. In the first quarter of 2020 their operating profits – earnings before interest and tax – were $1.6 billion, but in the same quarter of 2021 reached $27.1 billion. In 2020 the full-year profit of these carriers was around $25.4 billion, but 2021 it is likely to be an unprecedent $100 billion (Drewry, 2021). And this at a time of pandemic-related disruptions, congestion at ports and a persistent shortage of containers.

---

[1] TIM Market Intelligence Initiative Global Ocean Transport.
Overview & Methodology: TIM Consult are operating the Market Intelligence Initiative (MII) in global ocean transport (Full Container Load and Less Than Container Load) in support of a Community (consortium) of world-class enterprises (shippers only). The analyses cover ocean transport on more than 12,000 port pairs, pre- and on-carriage (all modes) and door-door-transport. The benchmarking as well as the monitoring of freight indices and service levels is updated on a monthly, quarterly, and annual basis. All input data is provided by shippers and represents actual agreements and volume allocations. No unnegotiated or not actually allocated rate information is included. Continuous data input is equivalent to approximately five per cent of world container transport. Data input is carefully cleansed by an expert team plus all strategic and operative drivers of rate and service levels as much as procurement performance clarified. The analyses and assessment of shippers' agreements are conducted by accurate segmentation (by box type, box size, port pair, process setup) and harmonization (normalization), taking into account all cost and service level drivers in full transparency. The rate benchmarking and the index information provided to UNCTAD are given on gate-in-gate-out level including all ocean transport-related charges and surcharges. Not included are pre- and on-carriage as much as classical administrative services of forwarders (customs clearance, booking and invoice control fees, etc.). MII members range from 1,000 TEU to 500,000 TEU per year. www.timconsult.com.

Increased earnings have encouraged carrier to order new ships. At the beginning of 2021 the orderbook for container ships was similar to that in 2018. As noted in chapter 2, the surge in new orders was also prompted by low prices for new, larger vessels and by the availability of ship financing.

Following the 2008–2009 financial crisis there was a similar rush in orders such that the container ship order book represented about 60 per cent of the global fleet, and new vessels started entering the market only a year after the crisis, leading to overcapacity and low freight rates. This is unlikely to happen now. Indeed the new ships are still unlikely to meet the demand. In recent years, shipping companies were faced with low earnings and uncertainties about complying with new IMO emission requirements, so had postponed placing orders (FitchRatings, 2021a). As it usually takes two to three years between the placement of vessel orders and delivery, the supply-demand imbalance is unlikely to be resolved in the short term so rates should remain high.

Indeed even the arrival of new ships may not be enough to reduce and stabilize container freight rates. Global freight rates will remain high until shipping supply-chain disruptions are unblocked and back to normal, and port constraints and terminal efficiencies are tackled (Hellenic Shipping News, 2021a). This would entail investing in new solutions, including infrastructure, freight technology and digitalization, and trade facilitation measures.

Moreover, even when they have new capacity, container lines faced with prolonged port congestion and closures may take capacity out of the system – keeping freight rates high. It can be argued that port congestion on the United States West Coast was initially caused by carriers responding to increased demand by inserting more capacity – but ports were then unable to handle the resulting surge. Moreover, despite recent improvements, overall port performance remains the lowest it has been in ten years of records (Global Maritime Hub, 2021).

All the above suggests that high freight rates may be sustained in both short and medium terms. This could have lasting effects on trade and global supply chains. By end of 2020 and early 2021, Europe was facing shortages of consumer goods imported from Asia – from home furnishings, bicycles and sports to children's toys and dried fruits. Some companies have stopped exporting to certain locations while others have been looking to shorten their supply chains by looking for goods or raw materials from nearer locations (Financial Times, 2021).

Another example is Viet Nam's exports of pepper. According to the Viet Nam Pepper Association, higher logistics costs have resulted in a loss of export markets. In 2020, for exports to the United States, the cost per 40-foot container was $2,000 to $3,000 but in the first six months of 2021 this had soared to an average $13,500. For exports to the European Union there was a corresponding increase, from $800-1,200 to $11,000. This caused importers to switch to pepper from Brazil; for the United States the shipping cost is only a third of that from Viet Nam and for the European Union only one tenth (Vietnamplus, 2021).

Shipping cost escalation, if sustained, would not only affect exports and imports, as well as production and consumer prices, but also the prospects for short- and medium-term economic recovery. A number of governments are worried about this, including China, Republic of Korea, United States, and Viet Nam, and have raised concerns about the shipping companies.[2] In China, faced with record highs in September 2020, the authorities had put pressure on carriers on the Transpacific routes for both pricing and capacity management and there were suggestions of setting a ceiling (Financial Times, 2020). In the Republic of Korea, to ensure that small and mid-sized shippers have access to capacity the government has announced a plan to subsidize shipping rates – a 20 per cent discount on freight rates and guaranteed shipping space if they sign long-term service contracts with domestic shipping lines (JOC.com, 2021).

B. DRY BULK FREIGHT RATES ALSO REACH HIGHS

In the first half of 2020, the demand shock from the COVID-19 pandemic added downward pressure to an overly supplied market and led to a drop in dry bulk shipping freight rates. The second half, in contrast, saw a rebound in demand for dry bulk cargo, particularly for iron ore and grain into China. Together with slower growth in the active fleet this pushed up freight rates. This was reflected in the Baltic Exchange Dry Index, which measures the cost of shipping various raw materials, such as coal, iron ore, cement, grain and fertiliser (figure 3.5). In February 2020 this stood at only 461 points but by July 2021 had reached 3,257 points.

---

[2] See: https://www.ft.com/content/a013548c-9038-4798-9b2e-f431c4eb2fba; https://splash247.com/chinese-authorities-say-there-needs-to-be-a-rates-ceiling-saade/; https://www.lloydsloadinglist.com/freight-directory/news/EU-shippers-call-for-box-line-competition-scrutiny/78198.htm#.YN3KJ0w6-Uk; and https://www.bloomberg.com/news/articles/2021-02-04/freight-cost-pain-intensifies-as-pandemic-rocks-ocean-shipping.

3. Freight rates, maritime transport costs and their impact on prices

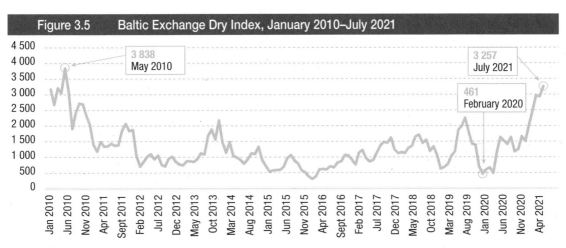

Figure 3.5   Baltic Exchange Dry Index, January 2010–July 2021

Source: UNCTAD, based on data from Clarkson *Shipping Intelligence Network*.

Freight rates were high through the first half of 2021 as a result of continuing higher demand, combined with fewer new vessel deliveries and increased scrapping activity. Rates were also affected by delays caused by port congestion. The number of vessels caught up in port congestion rose from 4 per cent of the fleet in the fourth quarter of 2020 to 5 per cent in the first quarter of 2021. This was mainly due to increases of exports of iron ore and grain products from Brazil which blocked up to 100 Capesize and Panamax vessels in Brazilian ports during February and March 2021 (Danish Ship Finance, 2021). The strength of the dry bulk market was good for carriers. In May 2020 the average monthly earnings of all bulkers were $4,894/day, but by June 2021 they were $27,275/day – the highest rates in a decade (figure 3.6).

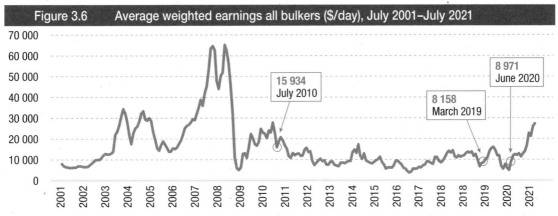

Figure 3.6   Average weighted earnings all bulkers ($/day), July 2001–July 2021

Source: UNCTAD, based on data from Clarkson *Shipping Intelligence Network*.

Looking ahead, dry bulk demand should continue to grow and the capacity should be manageable so rates are likely to remain high. The orderbook is only around 6 per cent of the existing fleet capacity, the lowest level in three decades (Clarksons Research, 2021b). Future freight rates will be largely determined by demand growth, particularly from China, but the market will also be affected by the ongoing energy transition and shifts in fuel mix choices. However, high freight rates could stimulate newbuild orders so that in the medium term, supply capacity could exceed demand.

C. TANKER FREIGHT RATES DIP TO THE LOWEST LEVELS EVER

In the first half of 2020, there was a surge in tanker freight rates, boosting profits for tanker shipping companies. In the second half of the year the COVID-19 impacts weakened demand and rates started to drop in an oversupplied market. By January 2021, oil tanker spot earnings were $5,237/day, and by July had fallen to $2,753/day, the lowest levels ever (figure 3.7). Given current low global demand and future uncertainties, short-term tanker freight rates will probably remain low.

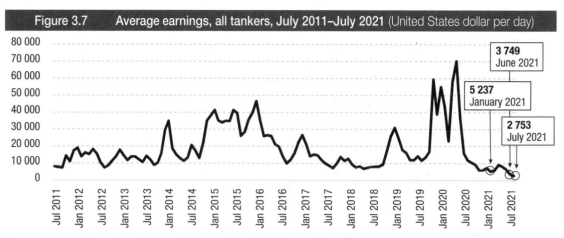

Figure 3.7 Average earnings, all tankers, July 2011–July 2021 (United States dollar per day)

Source: UNCTAD, based on data from Clarkson *Shipping Intelligence Network*.

D. ECONOMIC IMPACT OF HIGH CONTAINER FREIGHT RATES, PARTICULARLY IN SMALLER COUNTRIES

Containers offer efficient shipping services for a wide range of consumer and industrial commodities, including meats, beverages, textiles, and computers and by 2020 accounted for 17 per cent of the total volume of seaborne trade.[3] So, a surge in container freight rates will add to production costs which can feed through to consumer prices. This can slow national economies, particularly the structurally weak ones such as SIDS, LDCs, and landlocked developing countries (LLDCs) – whose consumption and production patterns are highly trade dependent. In 2019, for LDCs and LLDCs, merchandise imports made up 24 per cent of GDP, and for SIDS 58 per cent – compared with the global average of 21 per cent.[4]

1. High freight rates increase import and consumer prices, especially in SIDS

UNCTAD has simulated the impact of the current surge in container freight rates, concluding that at the global level import price levels will rise by 10.6 per cent, with an estimated one-year time lag (figure 3.8). This is an average for 200 economies for which data are available. The container freight rate surge refers to a 243 per cent increase in the CCFI between August 2020 and August 2021 and the simulation assumes that the levels in August 2021 will be sustained over the simulation period (technical note 1).

The impact is greatest in SIDS most of whose imports arrive by sea. In 2019, globally 27 per cent of total imports were seaborne, but for SIDS the proportion was 79 per cent.[5] As a result, the impact on their import prices is more than twice the global level, at 24 per cent. The situation is reversed for LLDCs: on average only one per cent of imports are transported by sea, so their import prices are simulated to increase by only 3.2 per cent.[6]

Increases in import prices also feed through to consumer prices. On average, for 198 economies for which data were available the global increase in prices between 2020 and 2023 is simulated at 1.5 per cent (figure 3.8). Consumer prices are less affected compared with import prices, due to the lower proportion of products that involve international shipping in the consumer basket. The level of increase also depends

---

[3] UNCTAD estimation.

[4] UNCTADstat (https://unctadstat.unctad.org/wds/TableViewer/tableView.aspx?ReportId=90759, accessed 26 July 2021). For the purposes of the analyses in this chapter, the definitions of LDC, LLDC, and SIDS follow the definitions of the Office of the High Representative for the Least Developed Countries, Landlocked Developing Countries and Small Island Developing States (UNOHRLLS) (https://www.un.org/ohrlls/content/profiles-ldcs, https://www.un.org/ohrlls/content/list-lldcs, https://www.un.org/ohrlls/content/list-sids, accessed 26 July 2021). The definition of SIDS includes Non-UN Members and Associate Members of the Regional Commissions.

[5] The share of maritime transport in SIDS total merchandise imports is calculated based on Comtrade Plus (https://comtrade.un.org/, accessed 16 June 2021) data for nine economies for which import value by mode of transport is available (i.e., Antigua and Barbuda, Belize, Comoros, Grenada, Guyana, Mauritius, São Tomé and Príncipe, Seychelles, and Suriname). The corresponding figure for non-SIDS is calculated based on Comtrade Plus data for 59 economies for which import value by mode of transport is available.

[6] The share of maritime transport in LLDC total merchandise imports is calculated based on Comtrade Plus data for 12 economies for which import value by mode of transport is available (i.e., Armenia, Azerbaijan, Botswana, Eswatini, Kyrgyzstan, Lao People's Democratic Republic, Mongolia, North Macedonia, Plurinational State of Bolivia, Republic of Moldova, Rwanda, and Zambia).

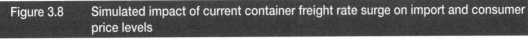

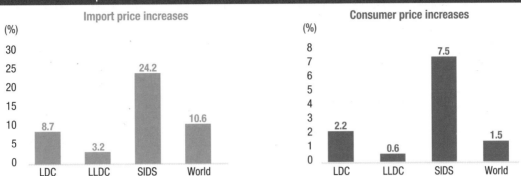

Figure 3.8 Simulated impact of current container freight rate surge on import and consumer price levels

Sources: UNCTAD calculations based on data provided by Clarksons Research, *Shipping Intelligence Network* (accessed 2 September 2021), the IMF, *International Financial Statistics* and *Direction of Trade Statistics* (accessed 1 June 2021), UNCTADstat (accessed 1-2 June 2021), and the World Bank, *World Integrated Trade Solution* (accessed 2 June 2021) and *Commodity Price Data* (The Pink Sheet, accessed 23 August 2021).

Note: Scenario with a 243 per cent freight rate increase compared to no freight rate increase (i.e., same freight rate level as August 2020) as a percentage of the import or consumer price level. The impacts of the container freight rate surge on prices are based on a 243 per cent increase in the CCFI between August 2020 and August 2021. See technical note 1 for the detail of the methodology.

on the extent to which wholesalers and retailers pass on the price increases; concerned about market share they may choose to absorb the import price increases by reducing their profits.[7]

In SIDS, the simulated increase is higher than the global average, at 7.5 per cent, because of their dependence on imports. The increase is also higher in LDCs than the global average at 2.2 per cent, partially because in high-inflation economies[8] firms tend to assume that increases in import prices will be persistent, and respond by increasing their prices.[9] In LLDCs, the increase in consumer prices is lower, at 0.6 per cent, owing to their limited dependence on maritime transport for imports.

2. Variations in price impacts across economies and types of goods

The adverse impacts of higher freight prices are not limited to SIDS and LDCs. Many other countries could see significant increases in consumer prices – ranging from 1.2 per cent in Brazil to 4.2 per cent increase in Slovakia (figure 3.9). It should be noted, however, that the simulation is limited to 27 European Union countries and 16 other major countries because it requires detailed information on sectoral-level input-output structures. The simulation assumes that all current freight increases and the corresponding increases in production costs are fully passed to consumers – with no change in other value-added components of production costs, such as wages and salaries (technical note 2).

The impact is generally greater in smaller economies. Thus, in Estonia consumer prices would rise by 3.7 per cent and in Lithuania by 3.9 per cent compared with only 1.2 per cent in the United States and 1.4 per cent in China. This partly reflects their greater 'import openness' – the ratio of imports to GDP – which is typically higher in smaller economies – 55 per cent in Lithuania and 60 per cent in Estonia, compared with 11 per cent in the United States and 15 per cent in China. Smaller economies are also likely to have a higher proportion of intermediate imported goods such as raw materials and components used for domestic production of consumer goods and services – 16 per cent in Lithuania and Estonia, compared with only 4 per cent in China and the United States.

---

[7] An empirical literature on exchange rate pass-through provides evidence that the low sensitivity of consumer prices to import price and exchange rate fluctuations can be explained by "double marginalization", wherein local wholesalers and retailers reduce their margins in response to exchange rate depreciations and import price increases to maintain market share at the retail level (Campa and Goldberg, 2010, and Hellerstein, 2008).

[8] Consumer price inflation in LDCs recorded 22.4 per cent in 2020, while the global inflation rate was 2.8 per cent (excluding the Bolivarian Republic of Venezuela due to its exceptionally high rate of inflation) according to UNCTADstat (https://unctadstat.unctad.org/wds/TableViewer/tableView.aspx?ReportId=37469, accessed 6 August 2021).

[9] An empirical literature on exchange rate pass-through provides evidence that emerging economies generally display higher sensitivity of domestic prices to exchange rate and import price fluctuations than developed countries, and the degrees of price sensitivity are affected by inflation rate levels and monetary policy credibility (Schmidt-Hebbel and Tapia, 2002; Choudhri and Hakura, 2006; McCarthy, 2007; Reyes, 2007; World Bank, 2014; Ha et al., 2020). The rationale for the correlation between price sensitivity and inflation is provided by the Taylor's hypothesis that firms in a higher and persistent inflation environment perceive exchange rate fluctuations to be more persistent and respond via price-adjustments (Taylor, 2000; Ca' Zorzi, et al., 2007).

Figure 3.9 Simulated impacts of the container freight rate surge on consumer price levels, by country and by product

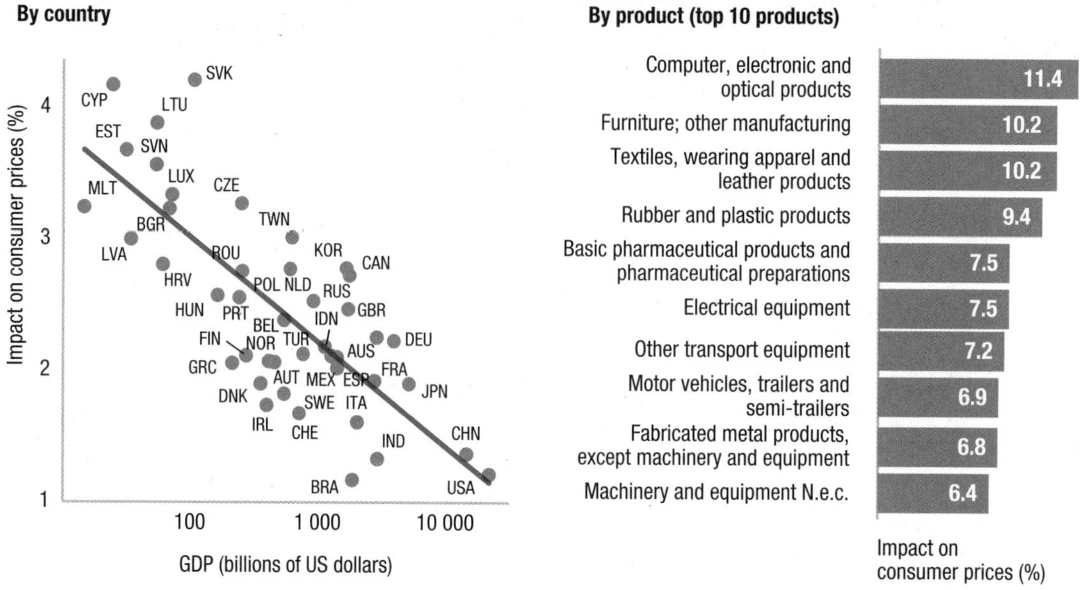

By country

(Scatter plot: Impact on consumer prices (%) vs GDP (billions of US dollars), log scale from 100 to 10 000. Countries plotted include SVK, CYP, LTU, EST, SVN, MLT, LUX, CZE, BGR, TWN, LVA, ROU, KOR, CAN, HRV, POL, NLD, RUS, HUN, PRT, BEL, IDN, GBR, FIN, NOR, TUR, AUS, DEU, GRC, AUT, MEX, ESP, FRA, DNK, SWE, ITA, JPN, IRL, CHE, IND, CHN, BRA, USA. A downward-sloping trend line shows higher impact on smaller economies.)

By product (top 10 products) — Impact on consumer prices (%)

Product	Impact (%)
Computer, electronic and optical products	11.4
Furniture; other manufacturing	10.2
Textiles, wearing apparel and leather products	10.2
Rubber and plastic products	9.4
Basic pharmaceutical products and pharmaceutical preparations	7.5
Electrical equipment	7.5
Other transport equipment	7.2
Motor vehicles, trailers and semi-trailers	6.9
Fabricated metal products, except machinery and equipment	6.8
Machinery and equipment N.e.c.	6.4

Sources: UNCTAD calculations based on the WIOD (accessed 7–8 June 2021) developed by Timmer et al., 2015, Clarksons Research, *Shipping Intelligence Network* (accessed 2 September 2021), UNCTADstat (accessed 24 June 2021), and the Centre d'Études Prospectives et d'Informations Internationales, *Gravity Database* (accessed 21 May 2021).

Note: The impacts of the container freight rate surge on prices are based on a 243 per cent increase in the CCFI between August 2020 and August 2021. The simulated impacts on price levels are long-term impacts, i.e., the simulation assumes that the current container freight rate surge and the corresponding increases in production costs are fully passed to consumers. See technical note 2 for the detail of the methodology.

Higher freight rates have a greater impact on the consumer prices of some goods than others, notably those which are more highly integrated into global supply chains, such as computers, and electronic and optical products (figure 3.9).[10] These often have to be shipped from East Asia towards consumption markets in the West with correspondingly higher shipping costs. For these goods, international shipping costs account for 2.6 per cent of the consumer price, compared with 1.2 per cent on average for other goods.[11] Higher prices will make such goods less affordable, so reduce consumer welfare.

Other goods for which surging freight rates are likely to increase consumer prices include low-value-added items such as furniture and textiles, wearing apparel and leather products.[12] Production of these goods is often fragmented across low-wage economies remote from major consumer markets. For example, international shipping costs account for 2.2 per cent of the consumer price for furniture and 1.8 per cent for textiles, wearing apparel and leather products.

3. Impact on global production processes and costs

Besides the consumer goods and services, other products that are closely integrated into global supply chains will be affected by surging freight rates. This is the case, for example, for investment-related products – capital goods and services used to create fixed assets, such as construction and computer programming (figure 3.10, technical note 2). Capital goods are more dependent than non-capital goods on supplies from foreign countries (Lian et al., 2020).

---

[10] Asia-Pacific Economic Cooperation (APEC), 2021 identified three key global value chain (GVC) industries in the APEC region based on their high values of GVC-related trade. They are computer, electronic and optical equipment, chemicals, and motor vehicle, trailers and semi-trailers. Among these three industries, computer, electronic and optical equipment showed the highest GVC participation rate in the APEC region.

[11] World average figures based on the World Input-Output Database (WIOD) used for the simulation. For this calculation (and the following calculations for furniture and textiles, wearing apparel and leather products), international shipping costs refer to only direct shipping costs of the final products from producer countries to consumer countries, and do not include shipping costs to source intermediate goods (i.e., raw materials and parts and components) used in the production process of the final products.

[12] For the purpose of the present analysis, furniture refers to furniture and other manufacturing sectors (i.e., divisions 31 and 32 in International Standard Industrial Classification, Rev.4, https://unstats.un.org/unsd/publication/seriesm/seriesm_4rev4e.pdf, accessed 30 July 2021).

3. Freight rates, maritime transport costs and their impact on prices

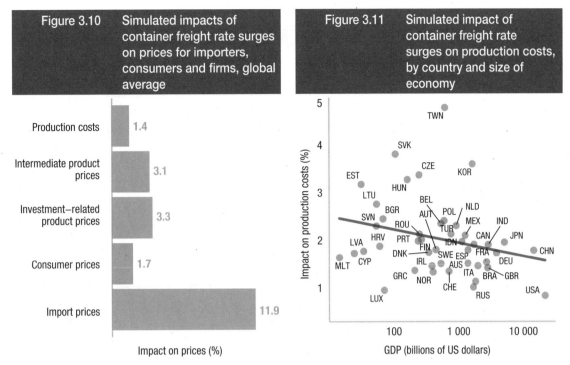

Figure 3.10 Simulated impacts of container freight rate surges on prices for importers, consumers and firms, global average

Figure 3.11 Simulated impact of container freight rate surges on production costs, by country and size of economy

Sources: UNCTAD calculations based on the WIOD (accessed 7-8 June 2021) developed by Timmer et al., 2015, Clarksons Research, *Shipping Intelligence Network* (accessed 2 September 2021), UNCTADstat (accessed 24 June 2021), and the Centre d'Études Prospectives and d'Informations Internationales, *Gravity Database* (accessed 21 May 2021).

Note: The impacts of the container freight rate surge on price levels are based on a 243 per cent increase in the CCFI between August 2020 and August 2021. The simulated impacts on price levels are long-term impacts, i.e., the simulation assumes that the current container freight rate surge and the corresponding increases in production costs are fully passed to final users (i.e., consumers and firms). See technical note 2 for the detail of the methodology.

Similarly, intermediate products are more strongly embedded in global supply chains than consumer products. These include raw materials, parts and components, and services used in production processes, such as banking and consultancy. For the dataset in the simulation, imported goods account for 14.6 per cent of total intermediate products used in domestic production processes, compared with 9.0 per cent for consumption products.

The impact is naturally lower for locally produced or assembled goods. Their production costs include not only the costs of intermediate products but also local value-added components such as labour. In the dataset used for the present simulation, globally these production factors account on average for 46 per cent of production costs. However, if the increase in prices triggers wage increases, this would increase the costs beyond those simulated.

Sustained increases in freight rates will cause greater increases in production costs in smaller economies and thus undermine their comparative advantages (figure 3.11). Smaller countries will also find it more difficult to move up the value chain if they face higher costs of importing high-technology machinery and industrial materials. This will hamper their efforts to achieve the Sustainable Development Goals.

4. Higher costs and maritime transport disruption threaten the recovery in global manufacturing

Manufacturers in the United States and Europe rely mainly on industrial supplies from China and other East Asian economies, so continued cost pressures, disruption and delays in containerized shipping will hinder production. The present analysis shows that a 10 per cent increase in container freight rates, along with supply chain disruptions, is expected to decrease industrial production in the United States and the euro area by more than 1 per cent cumulatively (figure 3.12, technical note 3).[13] In China, production is expected to decrease by 0.2 per cent. In the short to medium term these disturbances are likely to undermine recovery in manufacturing in major economies.

---

[13] In the present analysis, the euro area refers to 16 countries out of 19 euro area countries where all data are available for the simulation.

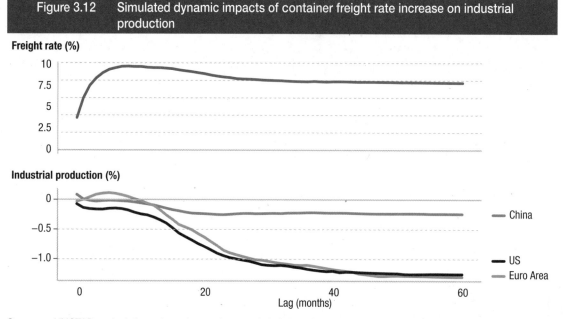

Figure 3.12 Simulated dynamic impacts of container freight rate increase on industrial production

Sources: UNCTAD calculations based on data provided by Clarksons Research, *Shipping Intelligence Network* (accessed 3 June 2021), the World Bank, *World Development Indicators* (accessed 10 June 2021), Bank for International Settlements, *Effective exchange rate indices* (accessed 10 June 2021), and Feldkircher et al., 2020 (accessed 10 June 2021).

Note: Global Vector Autoregression, consisting of 8 variables and 31 countries, is estimated using GVAR toolbox 2.0 (Smith and Galesi, 2014). Included endogenous variables for individual countries are the industrial production index, the consumer price index, the equity price index, the real effective exchange rate index, nominal short-term interest rates, and nominal long-term interest rates. Global variables are oil prices and container freight rates. See technical note 3 for the detail of the methodology.

As of July 2021, industrial production in the United States had recovered considerably from the decline caused by the COVID-19 pandemic in 2020, but remained below the pre-pandemic level despite strong consumer demand for goods. By early 2021, production in the United States had started to recover. Nevertheless compared with February 2020, by July 2021, industrial production was 0.1 per cent lower while real personal consumption expenditure on goods was 14.8 per cent higher.[14][15] These trends are consistent with the simulation for industrial production, suggesting that the container freight rate surge and the corresponding disruption in maritime transport are delaying a recovery in global manufacturing.

E. STRUCTURAL DETERMINANTS OF MARITIME TRANSPORT COSTS

As well as responding to global market factors such as strong shipping demand, limited supply and container shortages, maritime transport costs on specific routes are also determined by structural factors, including port infrastructure, trade facilitation measures and liner shipping connectivity. Indeed, compared with pandemic-induced fluctuations these can have a greater impact on transport costs and trade competitiveness in the long term. Improving these structural factors can mitigate future external shocks such as freight rate surges and maritime transport disruptions.

To investigate the structural determinants of maritime transport costs, UNCTAD has collaborated with the World Bank and Equitable Maritime Consulting to develop the Global Transport Costs Dataset for International Trade (GTCDIT).[16] This is a unique and comprehensive dataset disaggregated by mode of transport at commodity level (HS code 6-digit level). Transport costs are measured as differences between cost, insurance, and freight (CIF) values, and free on board (FOB) values. As of September 2021, data had been published for the year 2016. The dataset is currently being refined to improve data quality and add subsequent years.

---

[14] Based on data provided by the United States Board of Governors of the Federal Reserve System, Industrial Production and Capacity Utilization (https://www.federalreserve.gov/releases/g17/current/, accessed 27 September 2021).

[15] Based on data provided by the United States Bureau of Economic Analysis, Personal Income and Outlays (https://www.bea.gov/data/income-saving/personal-income, accessed 27 September2021).

[16] https://unctadstat.unctad.org/EN/TransportCost.html (accessed 24 June 2021).

3. Freight rates, maritime transport costs and their impact on prices

1. LDCs incur higher maritime transport costs

To capture overall trends in the GTCDIT, transport cost data have been aggregated for three importing country groups – LDCs, LLDCs and the world as a whole (figure 3.13). In 2016 the highest all-mode transport costs are for LLDCs at 11.6 per cent of FOB value, compared with 9.4 per cent for the world as a whole, and 9.7 per cent for LDCs. This is not surprising since many LLDCs are hampered by their geographical locations and depend on more expensive modes of transport such as air and road. For example the heatmap in figure 3.14 indicates especially high transport costs for Mongolia, Zimbabwe, Kyrgyzstan, the Republic of Moldova and Mali.

For maritime transport costs, figure 3.13 shows that the highest costs, at 7.6 per cent of FOB value, are in LDCs compared with a world average of 5.6 per cent. For LDCs, reducing maritime transport costs is a crucial development challenge as they rely on maritime shipping more frequently than others.

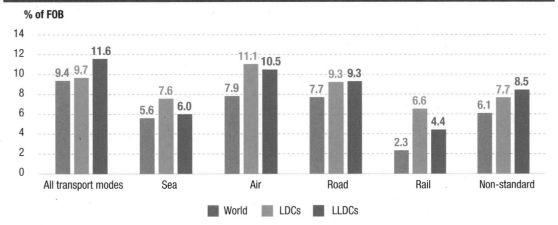

Figure 3.13 Transport costs for importing goods by transport mode, world, LDCs, and LLDCs, 2016, percentage of FOB value

Source: UNCTAD calculations based on the GTCDIT developed by UNCTAD, the World Bank, and Equitable Maritime Consulting (accessed 24 June 2021).

Note: Transport costs of each transport mode are aggregated by group of importing countries. The aggregation is the sum of transport costs over all commodities, importing countries in the respective importing country group, and trading partners, divided by the corresponding sum of the trade value (in FOB), for commodities and country pairs for which both transport costs and FOB values are available.

Figure 3.14 Transport costs heatmap for importing goods, all modes of transport, 2016, percentage of FOB value

Source: UNCTAD calculations based on the GTCDIT developed by UNCTAD, the World Bank, and Equitable Maritime Consulting (accessed 24 June 2021).

Note: Grey colour indicates countries where import transport costs data are not available.
Transport costs are aggregated by importing country. Importers' maritime transport costs are summed up over all commodities and trading partners and, divided by the corresponding sum of the trade value (in FOB), for commodities and country pairs for which both maritime transport costs and FOB values are available.

Maritime transport carried 56 per cent of the LDCs' total imports compared with a world average of 40 per cent.[17]

2. Better port infrastructure and trade facilitation would reduce maritime transport costs

The GTCDIT provides granular information on transport costs, which is useful to better understand the underlying relationships between these shipping costs and their determinants. This shows, for example, that, controlling for differences in product structure and local factors such as port infrastructure, the ad valorem maritime transport costs increase with the distance between trading partners, reflecting greater costs for fuel and crews. This relationship is visible in the granular data disaggregated at the commodity and bilateral country level (figure 3.15).[18] But it may not be evident in aggregated country level for average distance from trading partners. This is because some long-distance routes, such as between the United States and China, have larger volumes of trade that permit economies of scale, for example, by using larger vessels. Trade routes with longer distances and lower transport costs tend to have higher weights in the aggregation process.

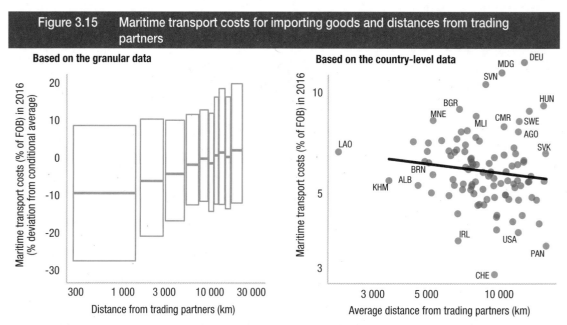

Figure 3.15 Maritime transport costs for importing goods and distances from trading partners

Source: UNCTAD calculations based on the GTCDIT developed by UNCTAD, the World Bank, and Equitable Maritime Consulting (accessed 24 June 2021).

Notes: Left-hand side: The granular data is the bilateral trade data at the HS code 6-digit level. Distances from trading partners are divided into ten quantile groups. The y-axis shows the percentage deviation of maritime transport costs from their conditional average based on commodities and trading partners (obtained as residuals from a regression of maritime transport costs (as percentage of the FOB value) on commodity dummies and trading partner dummies). The boxplot shows the 25th percentile (lower line), median (middle line), and the 75th percentile (upper line) of maritime transport costs in each quantile group.
Right-hand side: Importers' maritime transport costs are summed up over all commodities and trading partners and, divided by the corresponding sum of the trade value (in FOB), for commodities and country pairs for which both maritime transport costs and FOB values are available.

In ad valorem terms, maritime transport costs tend to be higher for smaller economies (figure 3.16). This may be due to the lack of liner shipping connectivity, the lower quality of port infrastructure, and inadequate trade facilitation measures. These countries would benefit from upgrading their ports to enable better shipping services, and permit larger vessels with shorter waiting times before entering ports. They

---

[17] The world average of the maritime transport share in terms of FOB value (i.e., 40.2 per cent) is lower than the maritime transport share in terms of volume (i.e., 85.9 per cent in 2016 according to Clarksons Research, *Shipping Intelligence Network*) indicating that goods transported by air and over land have on average a higher price than goods transported by sea.

[18] In the granular data, the elasticity of the maritime transport costs in ad valorem terms with respect to the distance is estimated at 0.059 after controlling commodity and trading partner fixed effects (and 0.028 without the fixed effects), and it is statistically different from zero at a significance level of 1 per cent. In contrast, in the country level data, the estimated elasticity is -0.091 and it is not statistically different from zero at a significance level of 10 per cent.

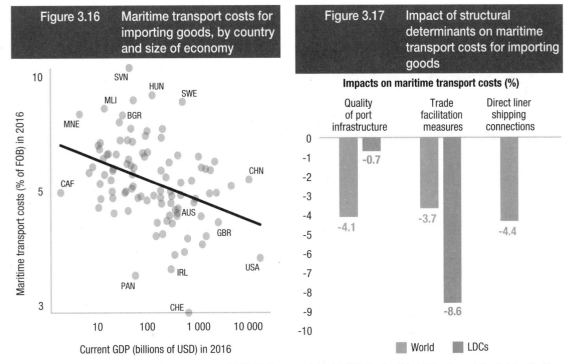

Sources: UNCTAD calculations based on the GTCDIT developed by UNCTAD, the World Bank, and Equitable Maritime Consulting (accessed 24 June 2021), *World Development Indicators* published by the World Bank (accessed 24 June 2021), *Global Competitiveness Index* published by the World Economic Forum (accessed 24 June 2021), *UN Global Survey on Digital and Sustainable Trade Facilitation* conducted by the UN Regional Commissions (accessed 24 June 2021), and a dataset provided by MDS Transmodal.

Notes: Figure 3.16: Maritime transport costs are aggregated by importing country. The aggregation is the sum of transport costs over all commodities and trading partners, divided by the sum of trade values (in FOB) over the corresponding commodities and trading partners, for commodities and country pairs where transport costs data are available.

Figure 3.17: The impact on maritime transport costs is the impact of improving each transport costs determinant from the 25th percentile to the 75th percentile. See technical note 4 for the detail of the methodology and the data sources.

would also benefit from introducing paperless systems for trade facilitation, as well as from more direct liner shipping connections to reduce the need for transhipping containers.

The consequence of improving these determinants – from their 25th percentiles to 75th percentiles – is illustrated in figure 3.17. Improving the quality of port infrastructure would reduce world average maritime transport costs by 4.1 per cent, better trade facilitation measures by 3.7 per cent, and better liner shipping connections by 4.4 per cent (technical note 4). In LDCs, the greatest benefits would come from better trade facilitation, with a decrease of 8.6 per cent compared with 0.7 per cent from better port infrastructure.[19]

It should be noted that these impacts are measured at border-to-border prices. As these transport costs determinants (quality of port infrastructure, trade facilitation measures, and liner shipping connection) would also reduce border-to-door transport costs, changes in total transport costs (door-to-door transport costs) can be expected to be higher than the changes in the border-to-border transport costs.

3. Trade imbalances produce asymmetric maritime transport costs, alleviated by economies of scale

Maritime transport costs are also affected by bilateral trade imbalances – especially for containerized trade. For sailings from high-demand to low-demand countries many vessels have to return with empty containers making shipping costs higher to cover part of the ballast sailing costs for the return journey.

This imbalance effect is confirmed in the data provided by TIM Consult (see section A.3). It is also evident in the GTCDIT dataset. Trade routes with trade imbalances on average have maritime transport costs 2.4 per cent higher for one direction than the other (figure 3.18). The greater the imbalance the greater

---

[19] Among trade facilitation measures, cross paperless trade and trade facilitation institution are estimated to have higher impacts in LDCs. Improving cross paperless trade and trade facilitation institution from the 25th percentile to 75th percentile is associated with a reduction in maritime transport costs by 8.8 per cent and 7.6 per cent, respectively.

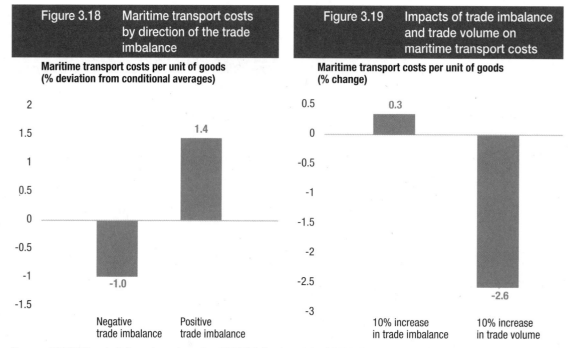

Figure 3.18 Maritime transport costs by direction of the trade imbalance

Figure 3.19 Impacts of trade imbalance and trade volume on maritime transport costs

Source: UNCTAD calculations based on the GTCDIT developed by UNCTAD, the World Bank, and Equitable Maritime Consulting (accessed 24 June 2021).

Notes: Figure 3.18: The figure shows the median of maritime transport costs in the sample of positive trade imbalances and the sample of negative trade imbalances. Maritime transport costs are percentage deviations from conditional averages based on commodities and bilateral country pairs (i.e., residuals from the regression of maritime transport costs (per unit of goods) on commodity dummies and bilateral country pair dummies). Differences in measurement unit of goods volume are controlled by the commodities dummies.

Figure 3.19: The figure shows the estimated elasticities (multiplied by 10) of maritime transport costs with respect to the trade (im)balance and the trade volume. See technical note 5 for the detail of the methodology and the data source.

the increase. Thus, if the imbalance increases by 10 per cent, maritime transport costs are expected to increase by 0.3 per cent (figure 3.19, technical note 5).[20]

The trade imbalance effect on maritime transport costs can be alleviated by other factors. For example, boosting cargo volumes to generate economies of scale could help cut maritime transport costs. The role of economies of scale effect in mitigating high transport costs is also confirmed when looking at the new transport costs dataset. An analysis based on this dataset shows that a 10 per cent increase in the trade volume is associated with a 2.6 per cent decrease in maritime transport costs (figure 3.19).

F. SUMMARY AND POLICY CONSIDERATIONS

Since late-2000 and into 2021, freight rates across containerized and dry bulk shipping markets have hit record highs, while tanker rates have plummeted. The surge in container rates in the second half of 2020 reflected higher-than-expected demand. As demand continued to surge, even an expansion of capacity was insufficient to constrain prices, because other supply-side factors came into play, including a global shortage of shipping containers, port congestion, delays, unreliable liner schedules, and increased fees and surcharges. Freight rates are expected to remain high – fuelled by continued strong demand against a background of growing supply uncertainty and concerns about the efficiency of transport systems and port operations.

The upward trajectory in freight rates has also raised questions about market behaviour and transparency in freight pricing – and about whether that situation has been exacerbated by greater market concentration.

The current surge in freight rates – if sustained – could have global economic impacts. The UNCTAD simulation suggests that it could increase global import price levels by 10.6 per cent, and consumer price levels by 1.5 per cent. The impact will be even greater in SIDS which could see import prices increase by 24 per cent and consumer prices by 7.5 per cent. In LDCs consumer price levels could increase by 2.2 per cent.

---

[20] In the quantitative analysis, the trade imbalance is defined as a ratio of the shipping value in one direction over the shipping value in the opposite direction.

Higher price increases are expected in important products. Globally, prices of computers are simulated to increase on average by 11 per cent, followed by 10 per cent increases in furniture and textiles, and a 7.5 per cent increase in pharmaceutical products. Some of these are low-value-added items produced in smaller economies which could face erosion of their comparative advantages.

Higher freight levels are also threatening to undermine a recovery in global manufacturing. In the short to medium term, a 10 per cent increase in container freight rates could lead to a cumulative contraction of around 1 per cent in industrial production in the United States and the euro area.

Over the longer term, maritime transport costs are also influenced by structural factors including port infrastructure quality, the trade facilitation environment, and shipping connectivity. There is potential for significant improvements that could reduce maritime transport costs by around 4 per cent.

If global trade is to flow more smoothly in future, and ports and maritime transport are to thrive and navigate through the historic disruption caused by the pandemic, this will require actions in some key policy areas, to:

- *Monitor markets* – To ensure a fair transparent and competitive commercial environment, governments will need to monitor freight rates, as well as fees and charges applied by carriers and port terminals. Policy makers should strengthen maritime transport competition authorities so that they can better understand market development and provide the requisite regulatory oversight (UNCTAD, 2021).

- *Share information and strengthen collaboration* – To enhance transport efficiency and operations there should be greater collaboration and sharing of data between various stakeholders along the maritime supply chain, including carriers, ports, inland transport providers, customs and shippers.

- *Analyse trends* – Relevant organizations, including UNCTAD, should continue to monitor trends in shipping markets, collect data and deepen their analysis of the structural determinants of transport costs. They can consider ways of cutting costs, enhancing efficiency and smoothing delivery of international maritime trade.

- *Upgrade ports* – To address congestion and ensure efficient and sustainable trade, port operations should be upgraded by improving infrastructure, and investing in new technology and digital solutions. Similar efforts should extend to trade facilitation to improve hinterland connectivity, particular for LDCs, SIDS and LLDCs.

- *Move up the value chain* – If smaller economies are to be more resilient to external shocks, including freight rate surges and maritime transport disruptions, they should be able to diversify by graduating to higher-value-added products.

REFERENCES

Asia-Pacific Economic Cooperation (2021). *APEC TiVA Initiative Report Two: Better Understanding Global Value Chains in the APEC Region*. Available at https://www.apec.org/Publications/2021/02/APEC-TiVA-Initiative-Report-Two---Better-Understanding-Global-Value-Chains-in-the-APEC-Region.

BIMCO (2020). Container Shipping: Capacity Management Key to Carriers' Profitability As Volumes Falter. Peter Sand. 09 September 2020. Available at https://www.bimco.org/news/market_analysis/2020/20200909_container_shipping.

Ca' Zorzi M, Hahn E, and Sánchez M (2007). Exchange Rate Pass-Through in Emerging Markets. The IUP *Journal of Monetary Economics*. 0(4): 84–102.

Choudhri EU and Hakura DS (2006). Exchange Rate Pass-Through to Domestic Prices: Does the Inflationary Environment Matter?. *Journal of International Money and Finance*. 25(4): 614–639.

Clarksons Research (2021a). *Container Intelligence Quarterly*. Second Quarter.

Clarksons Research (2021b). *Dry Bulk Trade Outlook*. Volume 27. April.

CNBC (2021). An 'aggressive' fight over containers is causing shipping costs to rocket by 300per cent. Weizhen Tan. January. Available at: https://www.cnbc.com/2021/01/22/shipping-container-shortage-is-causing-shipping-costs-to-rise.html.

Container xChange (2021). Demurrage & Detention Charges Benchmark 2021.

Danish Ship Finance (2021). Shipping Market Review. May.

Drewry (2021). Container Forecaster. Quarter 2. June.

Feldkircher M, Gruber T and Huber F (2020). International effects of a compression of euro area yield curves. *Journal of Banking & Finance*. 113105533.

Financial Times (2020). *China presses shipping line to rein in record freight rates*. David Keohane and Victor Mallet in Paris, George Steer in London. 17 November. Available at https://www.ft.com/content/a013548c-9038-4798-9b2e-f431c4eb2fba.

Financial Times (2021). *European retailers face goods shortages as shipping costs soar*. George Steer, Jonathan Eley and Valentina Romei in London. 31 Januray. Available at https://www.ft.com/content/40d23da5-c321-4b56-8ec7-551573a7a485.

Fitch Ratings (2021). Global Container Freight Rates to Moderate as Ship Orders Rise. *Fitch Wire*. 19 April. Available at https://www.fitchratings.com/research/corporate-finance/global-container-freight-rates-to-moderate-as-ship-orders-rise-19-04-2021.

Global Maritime Hub (2021). 5 reasons global shipping costs will continue to rise. Available at https://globalmaritimehub.com/5-reasons-global-shipping-costs-will-continue-to-rise.html.

Goldberg LS and Campa JM (2010). The Sensitivity of the CPI to Exchange Rates: Distribution Margins, Imported Inputs, and Trade Exposure. *The Review of Economics and Statistics*. 92(2):392–407.

Ha J, Stocker MM, and Yilmazkuday H (2020). Inflation and Exchange Rate Pass-Through. *Journal of International Money and Finance*. 105(C).

Hellenic Shipping News (2021a). Global Container Freight Rates to Moderate as Ship Orders Rise. International Shipping News. 20 April. Available at https://www.hellenicshippingnews.com/global-container-freight-rates-to-moderate-as-ship-orders-rise/.

Hellenic Shipping News (2021b). Uptrend in container rates looms over 2021 contract negotiations. March. Available at https://www.hellenicshippingnews.com/uptrend-in-container-rates-looms-over-2021-contract-negotiations/.

Hellerstein R (2008). Who bears the cost of a change in the exchange rate? Pass-through accounting for the case of beer. *Journal of International Economics*. 76(1):14–32.

JOC.com (2021). South Korea to subsidize container rates, boost logistics investment, Keith Wallis, July. Available at https://www.joc.com/maritime-news/container-lines/south-korea-subsidize-container-rates-boost-logistics-investment_20210706.html.

Jonkeren O, Demirel E, Van Ommeren J, and Rietveld P (2011). Endogenous transport prices and trade imbalances. *Journal of Economic Geography*, 11(3), 509–527. Retrieved July 6, 2021, from http://www.jstor.org/stable/26162209.

Korinek J (2011). Clarifying Trade Costs in Maritime Transport. Available at https://one.oecd.org/document/TAD/TC/WP(2008)10/FINAL/en/pdf (accessed 6 July 2021).

Lian W, Novta N, Pugacheva E, Timmer Y, and Topalova P (2020). The Price of Capital Goods: A Driver of Investment Under Threat. *IMF Economic Review*. 68.

McCarthy J (2007). Pass-Through of Exchange Rates and Import Prices to Domestic Inflation in Some Industrialized Economies. *Eastern Economic Journal*. 33(4): 511–537.

Miller RE and Blair PD (2009). Input–Output Analysis: Foundations and Extensions, Second Edition. Cambridge University Press.

Pesaran MH, Schuermann T, and Weiner SM (2004). Modeling Regional Interdependencies Using a Global Error-Correcting Macroeconometric Model. *Journal of Business & Economic Statistics*, 22(2): 129–162.

Reyes J (2007). Exchange Rate Passthrough Effects and Inflation Targeting in Emerging Economies: What is the Relationship? *Review of International Economics*. 15(3): 538–559.

Schmidt-Hebbel K and Tapia M (2002). Monetary Policy Implementation and Results in Twenty Inflation-Targeting Countries. *Working Papers Central Bank of Chile*. 166.

Sea Intelligence (2021). Press Release. June 15th. Availle at https://www.sea-intelligence.com/images/press_docs/ss518/20210615_-_Sea-Intelligence_Sunday_Spotlight_518_Press_Release.pdf.

Sekine T (2006). Time-Varying Exchange Rate Pass-Through: Experiences of Some Industrial Countries. BIS Working Papers. 202.

Smith LV and Galesi A (2014). GVAR Toolbox 2.0. Available at https://sites.google.com/site/gvarmodelling/gvar-toolbox (Accessed 6 July 2021).

Tamamura C (2014). Concept of Price Analysis Model by Input-Output Table and its Application to Asia Table. *International Input-Output Analysis: Theory and Application*. Institute of Developing Economies, Japan External Trade Organization.

Taylor, JB (2000). Low Inflation, Pass-Through, and the Pricing Power of Firms. *European Economic Review*. 44(7): 1389–1408.

Timmer MP, Dietzenbacher E, Los B, Stehrer R and Vries GJ de (2015). An Illustrated User Guide to the World Input–Output Database: the Case of Global Automotive Production. *Review of International Economics*. 23(3):575–605.

UNCTAD (2021). Container Shipping in Times of COVID-19: Why Freight Rates Have Surged, and Implications For Policymakers. Policy Brief. No.84. April. Available at https://unctad.org/system/files/official-document/presspb2021d2_en.pdf.

Vietnamplus (2021). *Vietnam at risk of losing pepper export markets due to high freight costs*. July 23. Available at https://en.vietnamplus.vn/vietnam-at-risk-of-losing-pepper-export-markets-due-to-high-freight-costs/205190.vnp.

World Bank (2014). Exchange Rate Pass-Through and Inflation Trends in Developing Countries. *Global Economic Prospects*. June.

TECHNICAL NOTES

Technical note 1: Simulation of import/consumer price impacts (section D.1)

The analysis in section D.1 simulated the impacts of the current container freight rate surge on import and consumer price levels at the world level and for three country groupings, i.e., LDCs, LLDCs, and SIDS. The simulated price impacts are defined as percentage differences in import/consumer price levels in 2023 between the following two scenarios:

1. **Container freight rate surge scenario:** The level of the CCFI Composite Index in August 2021 (i.e., 3,027.91 points) is assumed to be sustained over the remaining simulation period (i.e., from September 2021 to December 2023).

2. **No container freight rate surge scenario:** The CCFI Composite Index is assumed to stay at the level observed before the freight rate surge (i.e., 884.02 points in August 2020) over the remaining simulation period (i.e., from September 2020 to December 2023).

Estimation of the elasticities

The regression in the present analysis extended the exchange rate pass-through equation in Goldberg and Campa, 2010 and Sekine, 2006, to add container freight rates as an explanatory variable and expand the country coverage to include small countries such as LDCs, LLDCs and SIDS. Given that only annual data are available for most of the small countries, the number of observations is significantly reduced for each country. To overcome the small sample size problem, the estimation is conducted at the world level and the country group level instead of at the individual country level, applying a panel data estimation.

The first difference of logarithm of import prices is regressed on country dummies and the first differences of logarithms of container freight rates, nominal effective exchange rates, foreign prices, GDP, commodity prices, and lagged variables:

$$\Delta \ln IPI_t^c = \alpha^c + \sum_{l=0}^{L} (\beta_{1,l} \Delta \ln CCFI_{t-l}^c + \beta_{2,l} \Delta \ln e_{t-l}^c + \beta_{3,l} \Delta \ln w_{t-l}^c + \beta_{4,l} \Delta \ln GDP_{t-l}^c + \beta_{5,l} \Delta \ln Com_{t-l}^c) + \sum_{l=1}^{L} \beta_{6,l} \Delta \ln IPI_{t-l}^c$$

where IPI_t^c is local currency import price index of country c in year t, α^c is country fixed effects (i.e., dummy variables for country c), $CCFI_{t-l}^c$ is container freight rates of country c (i.e., freight rates of the closest trade lane for country c, to be discussed below) in year $t-l$, e_{t-l}^c is the inverse of the nominal effective exchange rate of country c, w_{t-l}^c is foreign prices (i.e., a weighted average of consumer prices of trading partners) of country c, GDP_{t-l}^c is the real GDP of country c, and Com_{t-l}^c is global commodity prices in terms of country c's local currency unit. For the construction of $CCFI_{t-l}^c$, each country is matched with the closest trade lane from the 12 trade lanes covered in the CCFI. For example, a country in Sub-Saharan Africa region is matched with the CCFI China-South Africa Freight Index. For e_{t-l}^c, the inverse of the nominal effective exchange rate is used in the equation, so that an increase in this variable represents a currency depreciation.

With regard to the impact on consumer prices, the first difference of logarithm of consumer prices is regressed on country dummies and the first differences of logarithms of import prices, GDP, and lagged variables.

$$\Delta \ln CPI_t^c = \alpha^c + \sum_{l=0}^{L} (\gamma_{1,l} \Delta \ln IPI_{t-l}^c + \gamma_{2,l} \Delta \ln GDP_{t-l}^c) + \sum_{l=1}^{L} \gamma_{3,l} \Delta \ln CPI_{t-l}^c$$

where CPI_t^c is consumer price index of country c in year t.

The above equations are estimated by OLS based on annual panel data. The import price equation covers 200 economies from 2003 to 2019, and the consumer price equation covers 198 economies from 1981 to 2019. As the coefficients (βs and γs) are common to all economies, estimated elasticities can be interpreted as the world average (simple average). For the estimation at the country group level (i.e., LDCs, LLDCs, and SIDS), the estimation samples are restricted to the respective country groups. For the import price equation, the sample sizes are 44 economies for LDCs (out of 46 LDCs), 31 economies for LLDCs (out of 32 LLDCs), and 42 economies for SIDS (out of 58 SIDS). For the consumer price equation, the sample sizes are 43 economies for LDCs, 31 economies for LLDCs, and 42 economies for SIDS. Insignificant explanatory variables are dropped from the equations, and consequently the lag lengths became 1 year for most cases.

Simulation of the impacts

To simulate the impacts of the current container freight rate surge on import prices, the estimated elasticities of import prices with respect to container freight rates is multiplied by the difference in freight rate between the container freight rate surge scenario and the no container freight rate surge scenario:

$$\beta_{1,0} \left(\Delta \ln CCFI_{2020}^{Composite} - \Delta \ln CCFI_{2020}^{Composite*} \right) \left(\sum_{l=0}^{3} \beta_{6,1}^{l} + \beta_{6,2} + 2\beta_{6,1}\beta_{6,2} + \beta_{6,3} \right)$$

$$+ \left[\beta_{1,0} \left(\Delta \ln CCFI_{2021}^{Composite} - \Delta \ln CCFI_{2021}^{Composite*} \right) \right.$$

$$\left. + \beta_{1,1} \left(\Delta \ln CCFI_{2020}^{Composite} - \Delta \ln CCFI_{2020}^{Composite*} \right) \right] \left(\sum_{l=0}^{2} \beta_{6,1}^{l} + \beta_{6,2} \right)$$

$$+ \left[\beta_{1,0} \left(\Delta \ln CCFI_{2022}^{Composite} - \Delta \ln CCFI_{2022}^{Composite*} \right) + \beta_{1,1} \left(\Delta \ln CCFI_{2021}^{Composite} - \Delta \ln CCFI_{2021}^{Composite*} \right) \right.$$

$$\left. + \beta_{1,2} \left(\Delta \ln CCFI_{2020}^{Composite} - \Delta \ln CCFI_{2020}^{Composite*} \right) \right] (1 + \beta_{6,1})$$

$$+ \left[\beta_{1,1} \left(\Delta \ln CCFI_{2022}^{Composite} - \Delta \ln CCFI_{2022}^{Composite*} \right) + \beta_{1,2} \left(\Delta \ln CCFI_{2021}^{Composite} - \Delta \ln CCFI_{2021}^{Composite*} \right) \right.$$

$$\left. + \beta_{1,3} \left(\Delta \ln CCFI_{2020}^{Composite} - \Delta \ln CCFI_{2020}^{Composite*} \right) \right]$$

where $CCFI_t^{Composite}$ is CCFI Composite Index in year t under the container freight rate surge scenario, and $CCFI_t^{Composite*}$ is CCFI Composite Index in year t under the no container freight rate surge scenario. Actual simulation equations are simpler because insignificant variables are dropped from the estimation equations. In the simulation, the CCFI Composite Index (instead of individual freight indices used in the estimation) is used for container freight rates to simplify the calculations.

A corresponding equation for the consumer price simulation can be obtained by replacing $CCFI_t^{Composite}$, $CCFI_t^{Composite*}$, $\beta_{1,l}$, $\beta_{6,l}$ with IPI_t, IPI_t^*, $\gamma_{1,l}$, $\gamma_{3,l}$, respectively, where IPI_t is import price index at the world level (or LDC, LLDC, or SIDS) in year t under the container freight rate surge scenario, and IPI_t^* is import price index under the no container freight rate surge scenario. IPI_t and IPI_t^* are calculated during the process of applying the above equation for the import price simulation.

Data

Import prices, consumer prices, real GDP, container freight rates, and commodity prices

Unit value indices of imports are reported in the UNCTADstat database (https://unctadstat.unctad.org/wds/TableViewer/tableView.aspx?ReportId=184185, accessed 2 June 2021). Given that the reported unit value indices are denominated in US dollars, they are converted to local currency units using market exchange rates. Data on market exchange rates are retrieved from the IMF, *International Financial Statistics* (https://data.imf.org/?sk=4c514d48-b6ba-49ed-8ab9-52b0c1a0179b, accessed 1 June 2021) and UNCTADstat (https://unctadstat.unctad.org/wds/TableViewer/tableView.aspx?ReportId=117, accessed 1 June 2021). For the 19 Euro area countries, the unit value indices of imports are converted to the former local currency units (before the Euro) because the dataset for the present analysis starts from 2003, which is before the adoptions of the Euro in some countries (i.e., Slovenia adopted the Euro in 2007, followed by Cyprus and Malta in 2008, Slovakia in 2009, Estonia in 2011, Latvia in 2014 and Lithuania in 2015).

Consumer price indices (CPI) and real GDP are retrieved from UNCTADstat (https://unctadstat.unctad.org/wds/TableViewer/tableView.aspx?ReportId=37469 for CPI and https://unctadstat.unctad.org/wds/TableViewer/tableView.aspx?ReportId=96 for real GDP, accessed 2 June 2021). CCFI composite index and the individual freight indices for 12 trade lanes are sourced from Clarksons Research, *Shipping Intelligence Network* (accessed 2 September 2021).

Commodity prices for energy, non-energy and precious metals are reported in the World Bank, *Commodity Price Data* (The Pink Sheet, https://www.worldbank.org/en/research/commodity-markets, accessed 23 August 2021). A simple average of the three indices are converted to local currency units using market exchange rates above.

Nominal effective exchange rates and foreign prices

The nominal effective exchange rate indices and the foreign price indices are normalized to 100 in the first year (i.e., 2003 for the most countries but a later year for some countries), and extended to subsequent years using the following chained formulas based on a geometric weighted average of bilateral exchange rates/trading partners' consumer price indices with trade values (i.e., bilateral total trade values for nominal effective exchange rates and bilateral import values for foreign prices) as weights:

$$NEER_t^c / NEER_{t-1}^c = \prod_{p \neq c} \left(\frac{E_t^c / E_t^p}{E_{t-1}^c / E_{t-1}^p} \right)^{W_t^{c,p}}, \quad w_t^c / w_{t-1}^c = \prod_{p \neq c} \left(\frac{CPI_t^p}{CPI_{t-1}^p} \right)^{\mathbf{W}_t^{c,p}}$$

where $NEER_t^c$ is the nominal effective exchange rate index of country c in year t, E_t^c is the market exchange rate of country c's currency in US dollars, E_t^p is the market exchange rate of trading partner p's currency in US dollars, and $W_t^{c,p}$ is the total bilateral trade value (i.e., the sum of the bilateral export value and the bilateral import value) between country c and trading partner p. For the right-hand side equation, w_t^c is the foreign price index of country c in year t, CPI_t^p is the consumer price index of trading partner p, and $\mathbf{W}_t^{c,p}$ is the bilateral import value of country c from trading partner p.

An increase in the nominal effective exchange rate index represents an appreciation of the country c's currency. In the estimation, the inverse of the nominal effective exchange rate index is used, so that an increase in this variable represents a currency depreciation.

The total bilateral trade value (i.e., $W_t^{c,p}$) and the bilateral import value (i.e., $\mathbf{W}_t^{c,p}$) are the average of the data reported by country c and trading partner p. If only either country c's or trading partner p's data is available, only the available data is used. If both data are not available, the missing value is imputed by the average of the previous and next year's values. Data on bilateral trade values and bilateral import values are retrieved from the IMF, *Direction of Trade Statistics* (https://data.imf.org/?sk=9D6028D4-F14A-464C-A2F2-59B2CD424B85, accessed 1 June 2021) and the World Bank, *World Integrated Trade Solution* (https://wits.worldbank.org/, accessed 2 June 2021). The data on market exchange rates (i.e., E_t^c and E_t^p) is the same data used in the calculation of import prices in local currency units (i.e., sourced from the International Monetary Fund, *International Financial Statistics* and UNCTADstat). Also, the data on trading partners' consumer price indices (i.e., CPI_t^p) is the same data used as the dependent variable in the consumer price equation (i.e., sourced from UNCTADstat).

Technical note 2: Simulation of price and production cost impacts (section D.2 and D.3)

The analyses in section D.2 and D.3 simulated the impacts of the current container freight rate surge on prices for importers, consumers and firms at the country level. The simulated impacts are "long-term" impacts, i.e., the simulation assumes that the current container freight rate surge and the corresponding increases in production costs are fully passed to final users (i.e., consumers and firms), although other production costs components such as wages and salaries are assumed not to change. The simulated impacts are defined as percentage differences in price/production cost levels between the following two scenarios:

1. **Container freight rate surge scenario:** The level of the CCFI in August 2021 (i.e., 3,027.91 points) is assumed to be sustained in the long-term (i.e., until increases in production costs are fully passed to final users).

2. **No container freight rate surge scenario:** The CCFI is assumed to stay at the level observed before the freight rate surge (i.e., 884.02 points in August 2020) in the long-term.

Estimation of the elasticities

In the first step, elasticities of production costs at the country and product level are estimated by the price model of the input-output table (see Tamamura, 2014; and Miller and Blair, 2009):

$$\boldsymbol{\eta} = \Delta(\boldsymbol{B}^t [\boldsymbol{b} + \boldsymbol{v} + \boldsymbol{d}]) = \boldsymbol{B}^t \Delta \boldsymbol{b}$$

where $\boldsymbol{\eta}$ is a column vector whose element η_i^c represents an elasticity of the production cost of product i in country c with respect to freight rates, $\boldsymbol{B}^t = \{[\boldsymbol{I-A}]^{-1}\}^t$ is the Leontief inverse matrix, $\boldsymbol{I}$ is an identity matrix (i.e., a square matrix with ones on the diagonal and zeros elsewhere), $\boldsymbol{A} = (a_{j,i}^{p,c})$ is the technical coefficient matrix and its element $a_{j,i}^{p,c} = Z_{j,i}^{p,c} / X_i^c$ represents the share of the input of product j produced in country p into the production of product i in country c (i.e., $Z^{p,c}$) in the total input for the production of product i in country c (i.e., X_i^c), $\boldsymbol{b}$ is a column vector whose element $b_i^c = IntTTM_i^c / X_i^c$ represents the ratio of the international transport margins involved in the production of product i in country c (i.e., $IntTTM_i^c$) over the total input for the production of product i in country c (i.e., X_i^c), $\boldsymbol{v}$ is a column vector whose element $v_i^c = VA_i^c / X_i^c$ represents the ratio of the value added (i.e., labour costs and capital costs) involved in the production of product i in country c (i.e., VA_i^c) over the total input for the production of product i in country c (i.e., X_i^c), and $\boldsymbol{d}$ is a column vector whose element $d_i^c = \tau_i^c / X_i^c$ represents the ratio of the indirect taxes less subsidies (i.e., import tariffs) involved in the production of product i in country c (i.e., τ_i^c) over the total input for the production of product i in country c (i.e., X_i^c).

The difference operator Δ represents element by element difference of a matrix/vector induced by a one per cent increase in container freight rates. Among the four matrices/vectors in the equation, i.e., $\boldsymbol{B}^t$, $\boldsymbol{b}$, $\boldsymbol{v}$, and $\boldsymbol{d}$, only the shares of the international transport margins (i.e., $\boldsymbol{b}$) are assumed to change. The share of transport margins involved in the production of product i in country c (i.e., b_i^c) is assumed to increase by one per cent if all imported products used in the production of product i in country c (i.e., $Z_{j,i}^{p,c}$ for all j,p) are fully containerized. If some imports are partially containerized, the transport margins of these products are assumed to increase by the containerized ratio divided by 100. Therefore, the change in the share of the international transport margins is calculated by the following formula:

$$\Delta b_i^c = \sum_{j,p} \left[Z_{j,i}^{p,c} \times R\_IntTTM_j^{p,c} \times \frac{CR_j^{p,c}}{100} \right]$$

where $b_i^c = \sum_{j,p} [Z_{j,i}^{p,c} \times R\_IntTTM_j^{p,c}]$ is the share of international transport margins involved in the production of product i in country c, $R\_IntTTM_j^{p,c}$ is the ratio of the international transport margins of product j's import from country p to country c over the import value of product j from country p to country c, and $CR_j^{p,c}$ is the containerized ratio of product j's import from country p to country c. The containerized ratio is calculated by the following formula:

$$CR_j^{p,c} = \left. \sum_{h \in j} MIMP_h^{p,c} \, 1_{containerized}(h) \right/ \sum_{h \in j} IMP_h^{p,c}$$

where $MIMP_h^{p,c}$ is the maritime import value of commodity h (in product group j) from country p to country c, $IMP_h^{p,c}$ is the total import value of commodity h (in product group j) from country p to country c, and $1_{containeriezed}$ is an indicator function which equals to one if commodity h is containerized and zero otherwise. The commodity h is considered as containerized according to the definitions used in the OECD Maritime Transport Cost database (see Appendix Table II.3. in Korinek, 2011).

In the second step, the elasticity of the final user prices (i.e., prices for consumers and firms) at the country and product level are estimated by summing the elasticity of the production costs η_i^p (estimated above) and the increase in the international transport margins for importing the product:

$$\zeta_i^{p,c} = \eta_i^p + \Delta R\_IntTTM_i^{p,c} = \eta_i^p + R\_IntTTM_i^{p,c} \times \frac{CR_i^{p,c}}{100}$$

where $\zeta_i^{p,c}$ is the elasticity of the final user price of product i imported from country p to country c, η_i^p is the elasticity of production cost of product i in country p, and $\Delta R\_IntTTM_i^{p,c}$ is the change in the international transport margin ratio of product i's import from country p to country c induced by a one per cent increase in container freight rates. If product i is fully containerized, the international transport margin ratio is assumed to increase by 1 per cent. Otherwise, the international transport margin ratio is assumed to increase by the containerized ratio divided by 100 (i.e., $CR_i^{p,c}/100$).

In the final step, the elasticity of the final user price and the elasticity of the production cost at the country and product level are aggregated to the country or product level using the final demand amounts or output values as weights:

$$\zeta^c = \sum_{i,p} \zeta_i^{p,c} f_i^{p,c}, \quad \zeta_i = \sum_{c,p} \zeta_i^{p,c} f_i^{p,c}, \quad \zeta^{global} = \sum_{i,c,p} \zeta_i^{p,c} f_i^{p,c}, \quad \eta^c = \sum_i \eta_i^c X_i^c$$

where ζ^c is the aggregated elasticity of final user prices in country c, ζ_i is the global elasticity of the final user price of product i, ζ^{global} is the global level elasticity of final user prices, η^c is the aggregated elasticity of production costs in country c, $f_i^{p,c}$ is the final demand of country c for product i produced in country p, and X_i^c is output of product i in country c. If the final demand vector $f = (f_i^{p,c})$ is the consumption of country c, the elasticity of final user prices (i.e., ζ^c) becomes the elasticity of consumer prices. The elasticities of import prices, investment-related product prices, and intermediate product prices are calculated by replacing the final demand vector by the respective demand vector.

Simulation

The impacts of the current container freight rate surge on prices and production costs at the country or product level are calculated by multiplying the aggregated elasticities by the changes in the CCFI level between the two scenarios:

$$\zeta^c \times \left(\frac{CCFI_{long}^{Composite}}{CCFI_{long}^{Composite*}} \times Adj - 1 \right) := \zeta^c \times \left(\frac{CCFI_{August\ 2021}^{Composite}}{CCFI_{August\ 2020}^{Composite}} \times Adj - 1 \right)$$

where $CCFI_{long}^{Composite}$ is the level of the CCFI Composite Index in the "long-term" under the container freight rate surge scenario (i.e., 3027.91 points in August 2021), $CCFI_{long}^{Composite*}$ is the level of the CCFI Composite Index in the "long-term" under the no container freight rate surge scenario (i.e., 884.02 points in August 2020), and Adj is an adjustment factor to convert changes in the CCFI to changes in international transport margin. Adj is calibrated by aligning changes in total international transport margin implied by the current simulation with changes calculated from a regression analysis at macroeconomic level (i.e., total international transport margin is regressed on the CCFI, and the estimation result is used for the extrapolation). The aggregated elasticity of final user prices at the country level (i.e., ζ^c) is replaced by the elasticity at the product level (i.e., ζ_i), at the global level (i.e., ζ^{global}), or the elasticity of production costs at the country level (i.e., η^c) when impacts on product level final prices, global level final prices or country level production costs are calculated.

Data

The estimation of the elasticities of prices and production costs at the country or product level is mainly based on the World Input-Output Database (WIOD, http://www.wiod.org/home, accessed 7-8 June 2021) developed by Timmer et al., 2015. The WIOD covers 43 countries (i.e., 28 EU countries and 15 other major countries) and 56 sectors. The calculation of the containerized ratio is based on the bilateral trade data by transport mode (the GTCDIT) retrieved from the UNCTADstat (https://unctadstat.unctad.org/EN/TransportCost.html, accessed 24 June 2021). The data on CCFI Composite Index is sourced from Clarksons Research, *Shipping Intelligence Network* (accessed 2 September 2021).

Technical note 3: Simulation of dynamic impacts on industrial production (section D.4)

The analysis in section D.4 simulated the dynamic impacts of container freight rate increases on the industrial production in major economies. The simulated impacts are defined as cumulative changes in the level of the industrial production induced by an increase in container freight rates.

Estimation

The regression is based on the global vector autoregression (GVAR) model developed by Pesaran et al., 2004. The GVAR consists of a set of vector autoregression (VAR) models at the individual country level:

$$x_{i,t} = a_{i,0} + a_{i,1} t + \sum_{l=1}^{p_i} \Phi_{i,l} x_{i,t-l} + \Lambda_{i,0} x^*_{i,t} + \sum_{l=1}^{q_i} \Lambda_{i,l} x^*_{i,t-l} + \Psi_{i,0} \omega_t + \sum_{l=1}^{q_i} \Psi_{i,l} \omega_{t-l} + u_{i,t}$$

$$\omega_t = \mu_0 + \mu_1 t + \sum_{l=1}^{p_\omega} \Phi_{\omega,l} \omega_{t-l} + \sum_{l=1}^{q_\omega} \Lambda_{\omega,l} x^*_{\omega,t-l} + \eta_t$$

$$x^*_{i,t} = \sum_{j \neq i}^{N} w_{i,j} x_{j,t}, \quad x^*_{\omega,t} = \sum_{j=0}^{N} w_{\omega,j} x_{j,t}$$

where $x_{i,t} = (y_{i,t}, \pi_{i,t}, eq_{i,t}, er_{i,t}, sr_{i,t}, lr_{i,t})^t$ are the country-specific endogenous variables of country i in time t, $x^*_{i,t} = (y^*_{i,t}, \pi^*_{i,t}, eq^*_{i,t}, er^*_{i,t}, sr^*_{i,t}, lr^*_{i,t})^t$ are the foreign variables (i.e., weighted average of foreign countries' endogenous variables) for country i, $\omega_t = (p_t^{oil}, p_t^{freight})^t$ are the global variables common for all countries, $w_{i,j}$ is the weight on country j's endogenous variables for constructing country i's foreign variables such that $\sum_{j \neq i}^{N} w_{i,j} = 1$, $w_{\omega,j}$ is the weight on country j's endogenous variables for constructing feedback variables for the global variables such that $\sum_{j=0}^{N} w_{\omega,j} = 1$, and $u_{i,t}$ are cross sectionally weekly correlated error terms. $y_{i,t}$ is the industrial production, $\pi_{i,t}$ is the consumer inflation, $eq_{i,t}$ is the real equity price, $er_{i,t}$ is the real effective exchange rate, $sr_{i,t}$ is the nominal short-term interest rate, $lr_{i,t}$ is the nominal long-term interest rate, p_t^{oil} is the oil price, and $p_t^{freight}$ is the freight rate. All variables are in levels and, with the exception of the interest rates, in logarithmic transform. Data on industrial production and consumer prices are seasonally adjusted.

In the country i's VAR model, $a_{i,0}$ is the intercept term, $a_{i,1}$ is the coefficient on the time trend term, $\Phi_{i,l}$ is the matrix of coefficients on the lagged endogenous variables, $\Lambda_{i,0}$ is the matrix of coefficients on the contemporaneous foreign variables, $\Lambda_{i,l}$ is the matrix of coefficients on the lagged foreign variables, $\Psi_{i,0}$ is the matrix of coefficients on the contemporaneous global variables, and $\Psi_{i,l}$ is the matrix of coefficients on the lagged global variables. In the VAR model for the global variables (i.e., the dominant unit model with the feedback effects), μ_0 is the intercept term, μ_1 is the coefficient on the time trend, $\Phi_{\omega,l}$ is the matrix of coefficients on the lagged global variables, and $\Lambda_{\omega,l}$ is the matrix of coefficients on the lagged feedback variables. The lag orders in the individual countries' VAR models and the dominant unit model (i.e., p_i, q_i, p_ω, and q_ω) are determined by the Akaike Information Criterion (AIC). The individual countries' VAR models and the dominant unit model are estimated using the GVAR toolbox 2.0 (Smith and Galesi, 2014).

Simulation

An impulse response analysis is conducted to simulate the impact of freight rate increases on the industrial production. The impact of the one standard deviation shock in freight rates is calculated by the generalized impulse response functions using the GVAR toolbox 2.0 (Smith and Galesi, 2014).

Data

The present analysis covers 31 major economies in the world (i.e., 24 countries in the EU-27 and 7 other major countries). The primary data source for the six endogenous variables (i.e., industrial production, consumer inflation, real equity prices, real effective exchange rate, nominal short-term interest rate, and nominal long-term exchange rate) is a dataset constructed by Feldkircher et al., 2020 (accessed 10 June 2021).

The other data sources used in the analysis are as follows: For the real effective exchange rates, the monthly real effective exchange rate indices (broad indices) calculated by the Bank for International

Settlements (https://www.bis.org/statistics/eer.htm, accessed 10 June) are used in the present analysis. For Container freight rates, the Containership Timecharter Rate Index is sourced from Clarksons Research, *Shipping Intelligence Network* (accessed 3 June 2021). For regional aggregation of the country level results and the construction of the feedback variables for global variables, current GDP based on purchasing power parity (PPP) is used as weights. The GDP data is sourced from the World Bank, *World Development Indicators* database (https://data.worldbank.org/indicator/NY.GDP.MKTP.PP.CD, accessed 10 June).

Technical note 4: Simulation of impacts of improving structural determinants on maritime transport costs (section E.2)

The analysis in section E.2 simulated the impacts of improving the structural determinants of maritime transport costs (i.e., the quality of port infrastructure, trade facilitation measures, and direct liner shipping connections) on maritime transport costs in ad valorem terms.

Estimation of the elasticities

The elasticities of maritime transport costs with respect to the structural determinants are estimated by the following panel regression:

$$\ln Cost_i^{c,p} = \alpha_i + \alpha^p + \beta x^c + \gamma \log(Dist^{c,p})$$

where $Cost_i^{c,p}$ is the maritime transport costs (per cent of FOB value) for importing commodity i (at the HS code 6-digit level) from country p to country c, α_i is the commodity fixed effects, α^p is the partner country (i.e., exporting country) fixed effects, x^c is a transport costs determinant of country c, and $Dist^{c,p}$ is the distance between country c and p. The country c's fixed effects are not included in the regression because they will cause the multicollinearity problem if they are included together with the transport costs determinants of country c. Only one transport costs determinant (i.e., either the quality of port infrastructure, trade facilitation measures, or the direct liner shipping connectivity) is included in the above equation at the same time to avoid the multicollinearity problem. The regression is run for each of the transport costs determinants to estimate the respective elasticity β.

When estimating the elasticities for the LDCs subsample, the equation is augmented to include an interaction term between the transport costs determinants and the dummy variable for the LDCs:

$$\ln Cost_i^{c,p} = \alpha_i + \alpha^p + \beta x^c + \delta(x^c \times Dum_{LDC}^c) + \gamma \log(Dist^{c,p})$$

where Dum_{LDC}^c is the dummy variable for LDCs and equals to one if country c is a LDC and zero otherwise. The elasticity of the maritime transport costs with respect to the transport costs determinants for LDCs is given by the sum of β and δ.

Simulation

To simulate the impacts of improving the structural determinants on maritime transport costs, the estimated elasticities are multiplied by the difference between the 25th percentile and the 75th percentile of the structural determinant: $\beta \times (x^{75th} - x^{25th})$, where x^{zth} is the zth (i.e., 75th or 25th) percentile of one of the transport costs determinants (i.e., the quality of port infrastructure, trade facilitation measures, or the direct liner shipping connectivity). In the simulation for the LDCs subsample, the formula is modified as follows: $(\beta + \delta) \times (x^{75th} - x^{25th})$.

Data

The maritime transport costs in 2016 at the commodity and bilateral country level are based on the Global Transport Costs Dataset for International Trade (GTCDIT, https://unctadstat.unctad.org/EN/TransportCost.html, accessed 24 June) developed by UNCTAD, the World Bank, and Equitable Maritime Consulting. The maritime transport costs in ad valorem terms are calculated by the following formula: $(CIF_i^{c,p} - FOB_i^{c,p}) / FOB_i^{c,p}$, where $CIF_i^{c,p}$ is the CIF value of commodity i's imports from country p to country c, and $FOB_i^{c,p}$ is the corresponding FOB value. The distance between the exporting country (i.e., country p) and the importing country (i.e., country c) is also recorded in GTCDIT.

The quality of port infrastructure is assessed in the Global Competitiveness Report published by the World Economic Forum. The score ranges from 1 (i.e., extremely underdeveloped) to 7 (i.e., well developed and efficient by international standards). The data for 2015-2016 are retrieved from the World Bank, TCdata360 (https://tcdata360.worldbank.org/indicators/IQ.WEF.PORT.XQ?country=BRA&indicator=1754&viz=line_chart&years=2007,2017, accessed 24 June). The data on trade facilitation measures are sourced from the *UN Global Survey on Digital and Sustainable Trade Facilitation* conducted by the UN Regional Commissions (https://www.untfsurvey.org/, accessed 24 June). The total trade facilitation score in 2015 is used in the analysis in the main text. The impacts of the five main individual scores (i.e., cross-border paperless trade, paperless trade, institutional arrangement and cooperation, formalities, and transparency) are also assessed and reported in relevant footnotes. For liner shipping connectivity, the number of directly connected countries in the liner shipping network (i.e., called degree centrality in the network analysis literature) is calculated based on a dataset provided by MDS Transmodal. Unlike the other two transport costs determinants, the logarithmic form is used for the estimation and simulation.

Technical note 5: Impacts of the trade imbalance and trade volume on maritime transport costs (section E.3)

Estimation of the elasticities

The analysis in section E.3 estimated the elasticity of maritime transport costs with respect to the trade (im)balance and the trade volume based on the following regression:

$$\ln Cost_i^{c,p} = \alpha_i + \alpha^{c,p} + \beta LBalance_i^{c,p} + \gamma \log(Volume_i^{c,p})$$

where $Cost_i^{c,p}$ is the maritime transport costs (per quantity unit of goods) for importing commodity i from country p to country c, α_i is the commodity fixed effects, $\alpha^{c,p}$ is the bilateral country pair fixed effects, $LBalance_i^{c,p}$ is the log of the trade balance of commodity i between country c and country p, and $Volume_i^{c,p}$ is the import volume of commodity i from country p to country c. The unit of the goods quantity used in the variables $Cost_i^{c,p}$ and $Volume_i^{c,p}$ is different by commodity. For example, the quantity of tomatoes is measured in kilograms while the quantity of textile wallcoverings is measured in square meters. The difference in the measurement unit is controlled by the commodity fixed effects α_i in the regression. Also, the impacts of the distance and the transport costs determinants analyzed in section E.2 (i.e., the quality of the port infrastructure, trade facilitation measures, and the direct liner shipping connections) are controled by the bilateral country pair fixed effects $\alpha^{c,p}$ in the present analysis.

The estimated elasticities, β and γ, are multiplied by 10 in figure 3.19. β represents the trade imbalance effect, and γ represents the economies of scale effect. It should be noted that the estimated economies of scale effect can be overestimated due to the reverse causality stemming from the trade promotion effect of low transport costs.

Data

All the variables used in the regression are based on the Global Transport Costs Dataset for International Trade (https://unctadstat.unctad.org/EN/TransportCost.html, accessed 24 June) developed by UNCTAD, the World Bank, and Equitable Maritime Consulting. The number of observations in the regression is 763,352 after selecting observations where the maritime trade value on the opposite direction is available.

The maritime transport costs per quantity unit of goods, $Cost_i^{c,p}$, are calculated by the following formula: $(CIF_i^{c,p} - FOB_i^{c,p}) / Volume_i^{c,p}$, where $CIF_i^{c,p}$ is the CIF value of commodity i's imports from country p to country c, and $FOB_i^{c,p}$ is the corresponding FOB value. The log of the trade balance, $LBalance_i^{c,p}$, is calculated by the following formula: $LBalance_i^{c,p} = \log(Value_i^{c,p}) - \log(Value_i^{p,c})$, where $Value_i^{c,p}$ is the import value (in terms of FOB) of commodity i from country p to country c, and $Value_i^{p,c}$ is the trade value of commodity i in the opposite direction (i.e., from country c to country p).

4

Key performance indicators for ports and the shipping fleet

This chapter provides key performance indicators based on a growing wealth of data derived from satellite tracking of vessels, shipping schedules, and port information platforms. Analysis of these data can help both users and providers of port and shipping services to compare progress and options and improve the efficiency of international maritime transport. The chapter has four sections.

A – Port calls – In early 2020, the pandemic initially resulted in a decline in ship arrivals, but there was a rebound in the second half of 2020 along with an increase in the median time that ships were spending in port. The advanced economies had higher volumes and lower turnaround times compared with smaller and less developed countries which suffered from diseconomies of scale and lower capacities. In Africa, those countries that had most container ship calls – Egypt and Morocco – also received larger vessels and had fast turnarounds.

B – Liner connectivity – There is a growing connectivity divide. Countries with low connectivity cannot generate the volume of trade that would encourage the frequent services they need to better connect to overseas markets. Among the 50 least-connected economies, 37 are small island economies.

C– Port performance – For container, dry-bulk, and tanker-port operations larger call sizes are associated with longer port stays, as it takes more time to load and unload greater volumes of cargo. However, if measured per ton or container of cargo, countries and ports with larger call sizes also record significantly better port performance. For large container ships the fastest average container handling speed is in Malaysia. For loading dry bulk cargo the highest productivity is in Australia, and for loading oil cargo it is in Angola.

D – Greenhouse gas emissions – Over the last decade, the world fleet has become more energy efficient. Nevertheless, there is continued growth in total GHG emissions, of which a high proportion is from container ships, particularly those that are older and less energy efficient. Ambitious measures will be needed to achieve the long-term goal of significantly reducing emissions.

Key performance indicators for ports and the shipping fleet

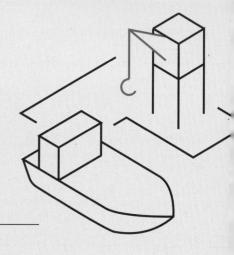

LINER SHIPPING CONNECTIVITY

The top 5 economies with the highest Liner Shipping Connectivity Index (LSCI) **are in Asia**

1 China	6 United States
2 Singapore	7 Spain
3 Republic of Korea	8 Netherlands
4 Malaysia	9 United Kingdom
5 Hong Kong, China	10 Belgium

The long-term trend in the distribution of the LSCI shows **a widening gap between the best and least connected countries**

18 of the 25 least connected economies and territories for which an LSCI has been generated **are islands**

PORT OPERATIONS

Larger ships and fewer port calls are two sides of the same coin

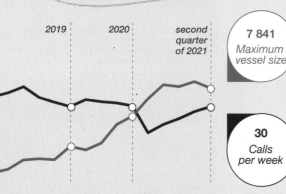

2019 2020 second quarter of 2021

7 841 Maximum vessel size

30 Calls per week

Ship sizes have increased faster than trade volumes and total deployed capacity

The fastest loading operation

Dry bulk carriers

Tonnes loaded per minute

- 48 Australia
- 28 Colombia
- 25 Brazil

Tankers

Tonnes loaded per minute

- 113 Angola
- 95 Qatar
- 90 Kuwait

CARBON DIOXIDE EMISSIONS

GHG emissions from shipping must be phased out to avoid the costs of not acting in the face of climate change

Decarbonization measures will have a greater impact on some countries than others, notably on SIDS or LDCs, which may need support to mitigate the increased maritime logistics costs

 The energy transition in maritime transport implies a major transformation of the industry

 In the process of decarbonizing shipping,

 maritime transport costs will increase,

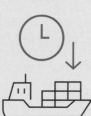 and average shipping speeds will decrease

 as a result, maritime logistics costs will go up

4. Key performance indicators for ports and the shipping fleet

A. PORT CALLS AND TURNAROUND TIMES

During the first six months of 2020, reflecting the pandemic-induced slump in demand for shipping and port services, the word's cargo-carrying ships as whole made fewer port calls (figure 4.1).[1] The second half of the year saw a rebound across all regions, albeit not to pre-pandemic levels. The highest number of ship arrivals were in Europe, East Asia, and South-East Asia (figure 4.2).

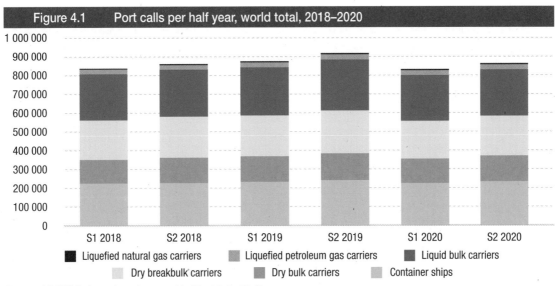

Figure 4.1 Port calls per half year, world total, 2018–2020

Source: UNCTAD, based on data provided by MarineTraffic.
Ships of 1,000 GT and above. Not including passenger and Ro/Ro ships.

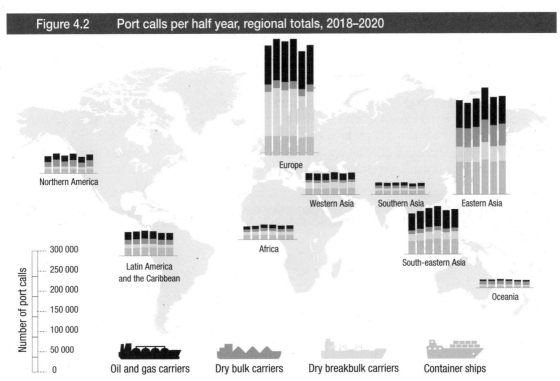

Figure 4.2 Port calls per half year, regional totals, 2018–2020

Source: UNCTAD, based on data provided MarineTraffic.
Note: Cargo carrying ships only, not including passenger ships and Ro/Ro vessels.

---

[1] UNCTAD secretariat calculations, based on data provided by MarineTraffic (www.marinetraffic.com). Aggregated figures are derived from the fusion of AIS information with port mapping intelligence by MarineTraffic, covering ships of 1,000 GT and above. For the computation of the turnaround times, passenger ships and RoRo ships are not included. Only arrivals have been taken into account to measure the number of port calls. Cases with less than ten arrivals or five distinct vessels on a country level per commercial market as segmented, are not included. The data will be updated semi-annually on UNCTAD's maritime statistics portal (http://stats.unctad.org/maritime).

During 2020, to contain the virus, terminal operators, authorities, and intermodal transport providers took steps to reduce social contact. However, this also slowed port operations so that vessels of all types had to spend more time in port (table 4.1). The greatest average increase in lengths of stay was for dry break bulk carriers whose general cargo operations tend to be more labour intensive and less automated. Moreover, when berth space is limited operators may prioritize scheduled container shipping calls or large dry bulk carriers over smaller vessels.

Table 4.1 Time in port, age, and vessel sizes, by vessel type, 2020, world total								
Vessel type	Median time in port (days), 2020	Median time in port, % change over 2019	Average size (GT) of vessels	Average age of vessels	Maximum size (GT) of vessels	Average cargo carrying capacity (dwt) per vessel	Maximum cargo carrying capacity (dwt) of vessels	Average container carrying capacity (TEU) per container ship
Container ships	0.71	2.3	38 308	14	237 200			3 543
Dry break bulk carriers	1.15	4.3	5 439	21	91 784	7 405	116 173	
Dry bulk carriers	2.07	2.7	32 146	14	204 014	57 453	404 389	
Liquefied natural gas carriers	1.12	0.8	95 270	12	168 189	74 229	156 000	
Liquefied petroleum gas carriers	1.04	3.0	10 826	15	59 229	12 164	64 220	
Wet bulk carriers	0.97	3.9	15 704	14	234 006	27 242	441 561	
All ships	**1.00**	**2.9**	**14 663**	**18**	**237 200**	**24 956**	**441 561**	**3 543**

Source: UNCTAD, based on data provided by MarineTraffic (https://www.marinetraffic.com).
Note: Ships of 1,000 GT and above. Not including passenger ships and Ro/Ro vessels.

Among the top 25 countries with the most container ship arrivals, the fastest median turnaround time was in Japan at 0.34 days, followed by Taiwan Province of China at 0.44 days, Hong Kong, China, at 0.52 days and China and Turkey both at 0.62 days (table 4.2). The longest average time in port was in the Russian Federation at 1.31 days, followed by Belgium at 1.04 days, the United States at 1.03 days and Indonesia at 0.99 days. For the container ships calling in its ports, the Russian Federation also recorded the highest average age and the smallest average size.

Figure 4.3 is a stylized map of port calls. It depicts container ship port calls per country, as well as the median time in port. Figure 4.4 does the same for container ship port calls and the maximum size of ship. Figure 4.5 and figure 4.6 zoom in on the same details for African countries. These figurative maps illustrate the importance of Asian economies. They also show that countries with more port calls tend to receive larger ships, while small island states can only accommodate fewer and smaller vessels.

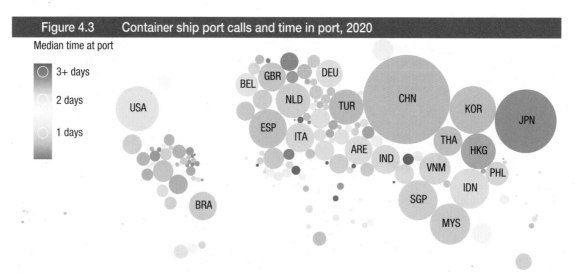

Figure 4.3 Container ship port calls and time in port, 2020

Source: UNCTAD, based on data provided by MarineTraffic.
Note: Ships of 1,000 GT and above. Labelled countries had more than 5,000 container ship port calls in 2020. For the complete table of all countries see http://stats.unctad.org/maritime.

Table 4.2 Port calls and median time spent in port, container ships, 2020, top 25 countries

Country	Number of arrivals	Median time in port (days)	Average age of vessels (years)	Average container carrying capacity (TEU) per vessel	Maximum container carrying capacity (TEU) of vessels
China	74 413	0.62	12	4 637	23 964
Japan	37 959	0.34	13	1 620	18 400
Republic of Korea	21 461	0.64	13	3 056	23 964
United States of America	18 866	1.03	14	5 347	22 000
Taiwan Province of China	16 621	0.44	14	2 665	23 964
Malaysia	15 875	0.80	14	3 706	23 756
Indonesia	15 019	0.99	14	1 509	14 855
Singapore	14 946	0.80	12	5 228	23 964
Spain	14 321	0.66	14	3 258	23 756
Hong Kong, China	11 976	0.52	13	3 637	23 964
Netherlands	11 595	0.80	14	2 942	23 964
Turkey	11 594	0.62	16	3 034	19 462
Viet Nam	9 587	0.90	13	1 966	18 400
Thailand	8 107	0.67	11	2 177	23 656
Italy	7 929	0.92	16	3 886	23 756
India	7 865	0.92	15	4 225	14 500
United Kingdom	7 834	0.73	15	3 465	23 964
United Arab Emirates	7 612	0.95	16	4 232	23 964
Brazil	7 609	0.77	10	5 877	12 200
Germany	7 139	0.98	13	4 442	23 964
Belgium	5 235	1.04	14	4 652	23 964
Philippines	5 181	0.89	15	1 858	6 622
Panama	4 467	0.69	12	4 139	14 414
Morocco	4 317	0.74	14	4 094	23 756
Russian Federation	4 184	1.31	18	1 509	9 400
Subtotal, top 25	**351 712**				
World total	**459 417**	**0.71**	**14**	**3 543**	**23 964**

Source: UNCTAD, based on data provided by MarineTraffic.
Note: Ships of 1000 GT and above. Ranked by number of port calls.
For the complete table of all countries, see http://stats.unctad.org/maritime.

Figure 4.4 Container ship port calls and maximum ship sizes, 2020

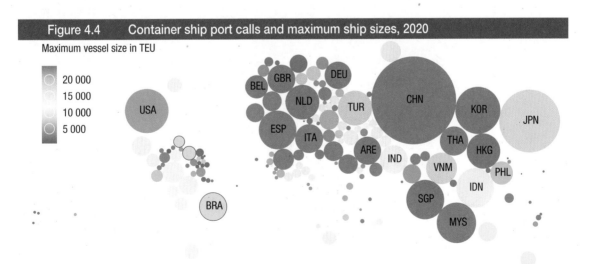

Source: UNCTAD, based on data provided by MarineTraffic.
Note: Ships of 1,000 GT and above. Labelled countries had more than 5,000 container ship port calls in 2020. For the complete table of all countries see http://stats.unctad.org/maritime.

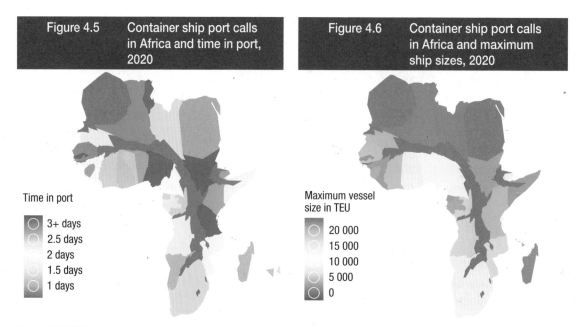

Figure 4.5 Container ship port calls in Africa and time in port, 2020

Figure 4.6 Container ship port calls in Africa and maximum ship sizes, 2020

Source: UNCTAD, based on data provided by MarineTraffic.
Note: Ships of 1,000 GT and above.

Source: UNCTAD, based on data provided by MarineTraffic.
Note: Ships of 1,000 GT and above.

The longest times in port are generally in Africa – notably in Nigeria, Sudan, and Tanzania – though Morocco is an exception with one of the world's shortest times.

Large ships with more cargo to be loaded or unloaded will normally require longer in port, though ports that can handle larger ships also tend to be more modern and better equipped, so can work more quickly and this is therefore a non-linear relationship (figure 4.7).

Some of the fastest turnarounds are in countries that have very few port calls and only receive ships with a few containers to be loaded and unloaded, so there is little congestion. However, at the other end of the scale, turnarounds are also fast in countries that have many port calls and can accommodate the largest container vessels. These ports benefit from economies of scale and investments in the latest technologies and infrastructure; their efficiency in turn attracts more vessels, further boosting the number of arrivals.

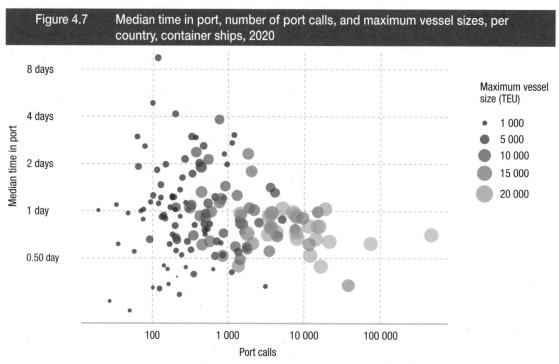

Figure 4.7 Median time in port, number of port calls, and maximum vessel sizes, per country, container ships, 2020

Source: UNCTAD, based on data provided by MarineTraffic. Both axes in logarithmic scale.
Note: Ships of 1,000 GT and above. For the complete table of all countries, see http://stats.unctad.org/maritime.

Countries in the middle of the distribution report a wide range of median times, reflecting differences in efficiency and other variables such as vessel age and cargo throughput.

B. LINER SHIPPING CONNECTIVITY

Since 2020, UNCTAD, in collaboration with MDS Transmodal, has reported quarterly values, at both port and country levels, for the Liner Shipping Connectivity Index (LSCI).[2] Countries with better liner shipping connectivity as reflected in the LSCI, generally have better access to overseas markets so can be more competitive (UNCTAD, 2017).

In the second quarter of 2021, the top-five most-connected economies, with the highest LSCIs, were in Asia – China, Singapore, Republic of Korea, Malaysia, and Hong Kong, China. These were followed by the United States and four European countries – Spain, Netherlands, United Kingdom, and Belgium (figure 4.8). In the four succeeding quarters, China widened its lead, while the United States saw a decline

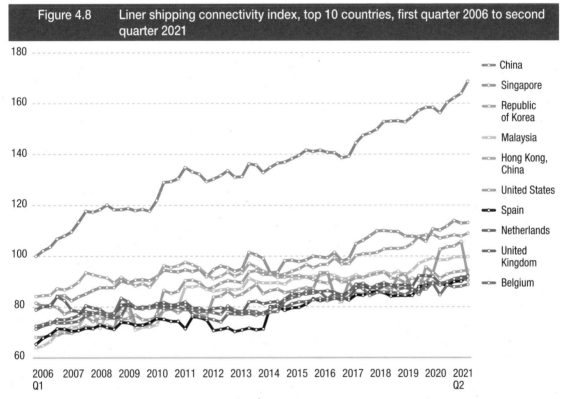

Figure 4.8 Liner shipping connectivity index, top 10 countries, first quarter 2006 to second quarter 2021

Source: UNCTAD, based on data provided by MDS Transmodal. For the complete data set for all countries see http://stats.unctad.org/LSCI.

---

[2] UNCTAD developed the Liner Shipping Connectivity Index (LSCI) in 2004. The basic concepts and major trends are presented and discussed in detail in (UNCTAD, 2017) and (MDST, 2020).

In 2019, the LSCI, in collaboration with MDS Transmodal (https://www.mdst.co.uk) was updated and improved, comprising additional country coverage including several SIDS, and incorporating one additional component, covering the number of countries that can be reached without the need for transhipment. The remaining five components, notably the number of companies that provide services, the number of services, the number of ships that call per month, the total annualized deployed container carrying capacity, and ship sizes, have remained unchanged. Applying the same methodology as for the country-level LSCI, UNCTAD has generated a new port Liner Shipping Connectivity Index.

Each of the six components of the port LSCI captures a key aspect of a connectivity.

(a) A high number of scheduled ship calls allows for a high service frequency for imports and exports.

(b) A high deployed total capacity allows shippers to trade large volumes of imports and exports.

(c) A high number of regular services from and to the port is associated with shipping options to reach different overseas markets.

(d) A high number of liner shipping companies that provide services is an indicator of the level of competition in the market.

(e) Large ship sizes are associated with economies of scale on the sea-leg and potentially lower transport costs.

(f) A high number of destination ports that can be reached without the need for transhipment is an indicator of fast and reliable direct connections to foreign markets.

Since 2020, the same methodology is applied on the country and the port level on a quarterly basis.

because of the inactivity in the second quarter of a trans-Pacific service of the 2M Alliance which had deployed ultra-large container carriers.

Of the 25 least-connected economies and territories for which an LSCI has been generated, 18 are islands whose LSCI scores have not significantly improved over the last 15 years. These are Anguilla, Antigua and Barbuda, Bermuda, Bonaire, Saint Eustatius and Saba, Cabo Verde, Cayman Islands, Christmas Island, Cook Islands, Micronesia, Montserrat, Niue, Norfolk Island, Palau, São Tomé and Príncipe, Saint Kitts and Nevis, Timor-Leste, Turks and Caicos Islands and Tuvalu. Among the bottom 25, two countries, Moldova and Paraguay, are landlocked so their LSCIs are determined by containerized river transport services. The remaining five economies are Albania, Democratic Republic of Congo, Eritrea, Gibraltar and Guinea-Bissau, whose seaborne trade is often handled by ports in neighbouring countries.

1. A growing connectivity divide

Over the period 2006–2021 the LSCI indicates a widening gap between the best- and least-connected countries, reflected in the dataset as an increase in the standard deviation, from 20 to 28. Over this period, China increased its LSCI by 69 per cent while many SIDS saw their LSCIs stagnate.

Among the 50 least-connected economies, 37 were small island economies. The exceptions were Bahamas, Jamaica and Mauritius which have high and growing LSCIs because they have developed into regional hubs, attracting transhipment of containerized trade for other countries. They can thus also offer their own importers and exporters better access to overseas markets (UNCTAD, 2021b).

Figure 4.9 depicts the LSCI at port level. Eight of the top ten ports were in Asia, led by Shanghai; the remaining two are in Europe – Rotterdam and Antwerp. The best-connected port in Latin America and the Caribbean was Cartagena, Colombia; in South Asia it was Colombo, Sri Lanka; in North America it was New York/New Jersey, United States; and in Africa it was Tanger Med, Morocco (figure 4.10).

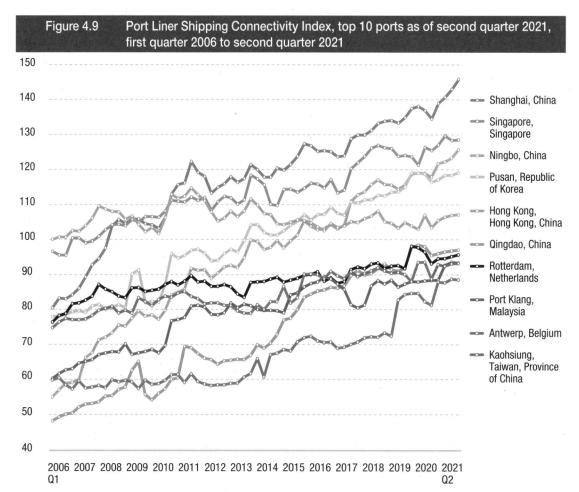

Figure 4.9 Port Liner Shipping Connectivity Index, top 10 ports as of second quarter 2021, first quarter 2006 to second quarter 2021

Source: UNCTAD, based on data provided by MDS Transmodal. For the complete data set for all ports see https://unctadstat.unctad.org/wds/TableViewer/tableView.aspx?ReportId=170026.

4. Key performance indicators for ports and the shipping fleet

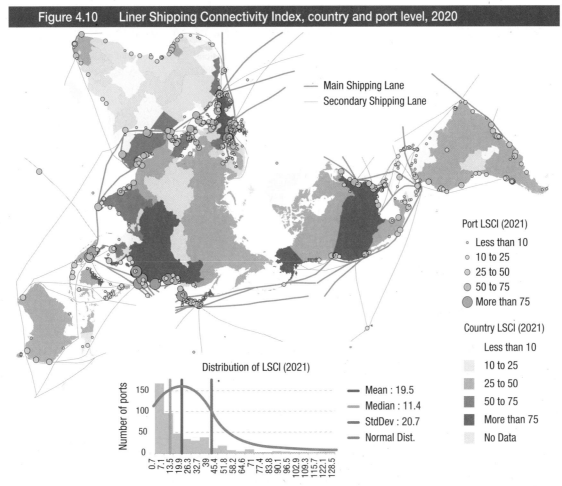

Figure 4.10 Liner Shipping Connectivity Index, country and port level, 2020

Source: Jean-Paul Rodrigue, Dept. of Global Studies & Geography, Hofstra University, based on data provided by UNCTAD. LSCI values are average of all 4 quarters of 2020.

2. Larger ships and fewer companies

To cater for growing demand, there are two main options. Carriers can either deploy more ships, and offer more services and direct connections, or they can deploy larger ships, or a combination of the two. In practice, over the last two decades, they have tended to use larger ships (figure 4.11).

The size of the largest ships has increased significantly, while the average number of companies has decreased. The outcome over this period was a 280 per cent increase in deployed capacity per company per country. Ship sizes have increased faster than trade volumes and total deployed capacity, so if ships are to remain fully loaded they will generally operate on fewer services. Between the first quarter of 2006 and the second quarter of 2021, the average capacity of the largest ship for each country increased by 176 per cent – from 2,836 to 7,841 TEU, while the average number of companies per country fell from 18 to 13.

Between the first quarter of 2006 and the second quarter of 2021 the capacity of the largest ships for each country increased by 155 per cent, to 23,963 TEU. In 2006, four countries had calls from more than 100 companies – Belgium, China, United Kingdom, and the United States. But by the second quarter of 2021, ports in China had services from only 93 companies, followed by Republic of Korea at 63 companies, the United States at 61, and Japan at 60.

Figure 4.12 Illustrates the trends in maximum vessel sizes and number of companies for selected countries from different global regions. Most countries have bigger ships and fewer companies. Among the countries covered in figure 4.12, between 2006 and 2021 the greatest growth in vessel size was in Chile, up by more than 300 per cent, from 3,430 to 14,300 TEU, while the greatest fall in number of companies was in Germany, from 97 to 38.

For the SIDS the situation is different. They generally offer limited and scattered markets so there is little justification for larger ships. The number of companies providing services for most SIDS has remained small, and there is little competition (see Samoa in figure 4.12).

Figure 4.11   Trends in global container ship deployment, first quarter 2006 to second quarter 2021

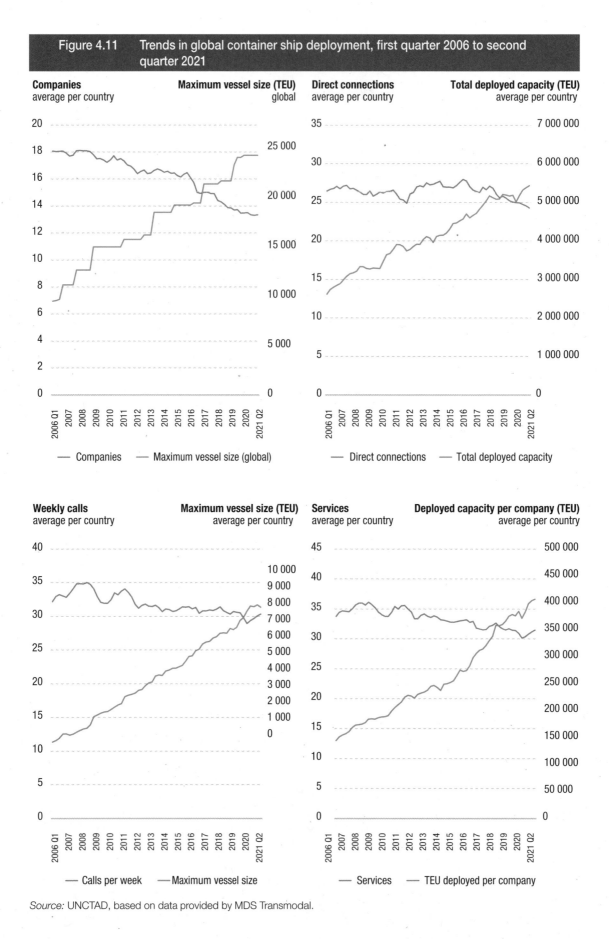

Source: UNCTAD, based on data provided by MDS Transmodal.

The relationship between total deployed container carrying capacities, ships sizes, and the number of companies in a market is further illustrated in figure 4.13. Moving vertically in the chart, for a given number of companies in a market, the total deployed capacity – how many containers can be carried to or from

4. Key performance indicators for ports and the shipping fleet

Figure 4.12 Trends in vessel sizes and number of companies providing services, selected countries, first quarter 2006 to second quarter 2021

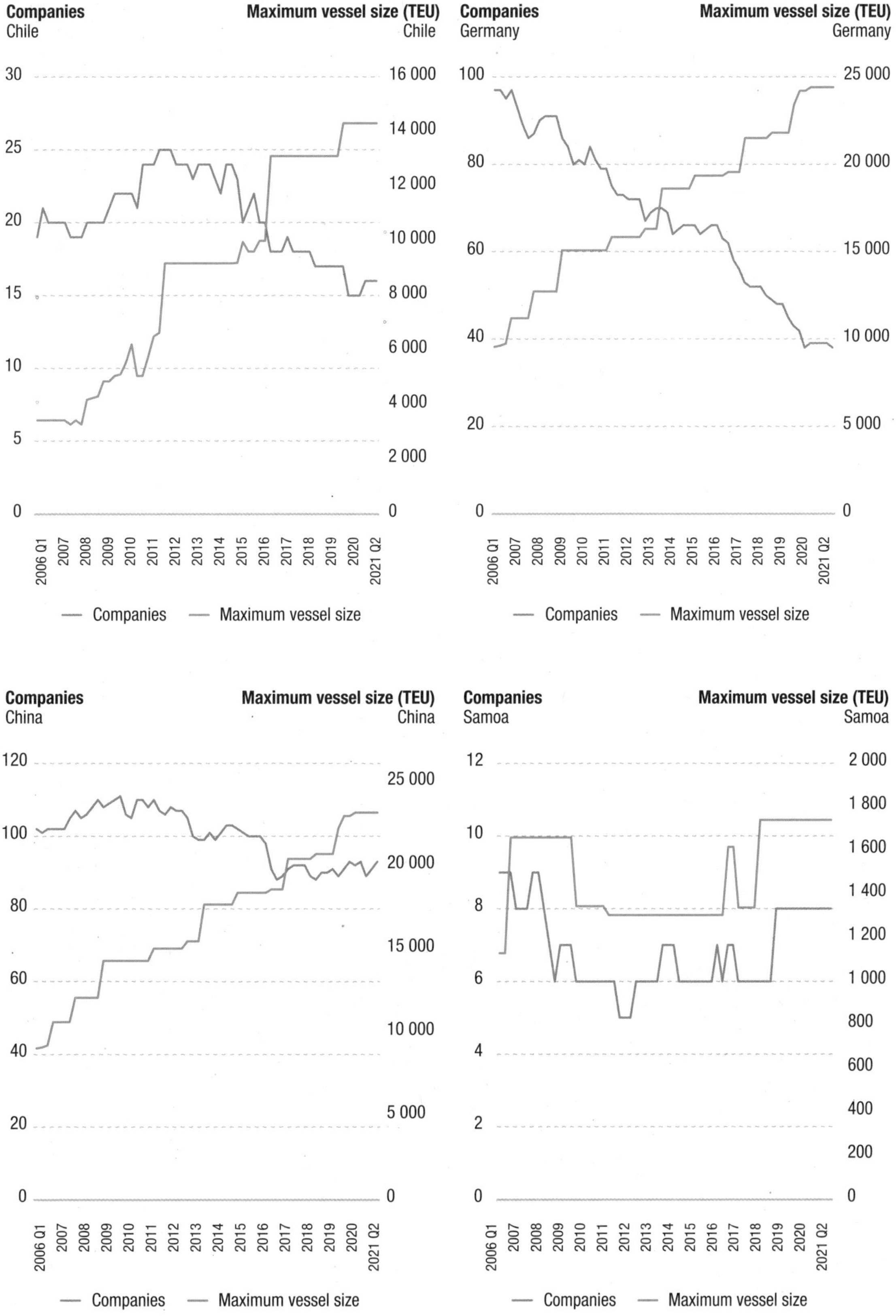

Source: UNCTAD, based on data provided by MDS Transmodal.

a country – increases with maximum vessel size. For each country, however, there is a trade-off between accommodating more companies or receiving larger ships: moving horizontally in the chart, for a given deployed capacity, the bigger ships are in countries with fewer companies in their markets.

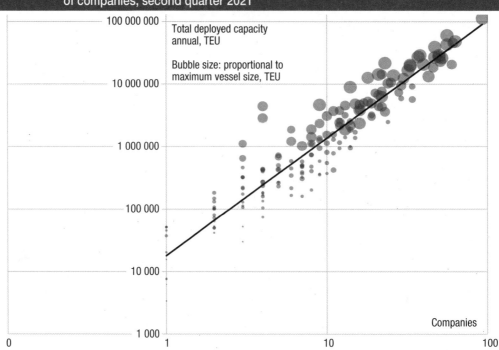

Figure 4.13 Relationship between maximum vessel sizes, deployed capacity, and the number of companies, second quarter 2021

Source: UNCTAD, based on data provided by MDS Transmodal.

3. Bilateral liner shipping connectivity

In addition to the country- and port-level LSCI, UNCTAD also produces a connectivity index for country pairs, the Liner Shipping Bilateral Connectivity Index (LSBCI).[3] Progress in the LSBCI, along with its five component indicators, is illustrated in figure 4.14. Since 2006, on average the LSBCI has increased but there have been a few disruptions – notably the global financial crisis of 2008, and the pandemic from 2020. The financial crisis had an almost immediate impact, but the pandemic impact came in waves – delivering a supply shock that then translated into a demand shock along with differences between countries in the local impact and propagation of the virus.

In addition to these disruptions, since the last quarter of 2018 the LSBCI has shown a downward trend which is more a consequence of ongoing structural transformations. One is the increase in ship size. Between 2006 and 2019 the maximum capacity component of the index more than trebled. Between 2014 and 2019 this was largely offset changes in the other four components, all of which have been declining.

These trends for the component indicators are interlinked. Companies that have invested in larger ships are aiming for economies of scale which should reduce unit costs. Other companies unable to make these investments, and to compete, will either withdraw from unprofitable routes or leave the industry altogether. This reduces the number of operators, which has been happening in all regions – in East Asia for the last seven years, but also in Latin America, and in Sub-Saharan Africa which in addition has fewer operators offering intra-regional connections.

With fewer companies, there are likely to be fewer direct connections. This is confirmed by the evolution of the transhipment component and consequently of the common direct component. Nevertheless, as direct connections are mainly on historical maritime routes the main adjusting variable on those routes is likely to be the number of competing companies.

Increasing ships size also affects the hosting capacity of ports especially those that have improved their infrastructure. This could explain the downward trend since 2017 for the frequency component which reflects the number of port-to-port connections between countries.

---

[3] The Liner Shipping Bilateral Connectivity Index (LSBCI), which is publicly available in its annual form at http://stats.unctad.org/lsbci, is made of five components: the number of transhipments needed to connect two countries (transhipment variable), the number of common direct connections between two countries (common direct variable), the number of port-to-port connections between two countries (frequency variable), the number of liner shipping companies operating between two countries (operators variable) and, the maximum ship size in TEU deployed between two countries (max. capacity variable). When no direct connection exists between two countries the latter three components correspond to connection (option) with the best (highest) value when taking the lowest connecting segment.

Figure 4.14 Liner Shipping Bilateral Connectivity Index (LSBCI) and its components, first quarter 2006 to second quarter 2021

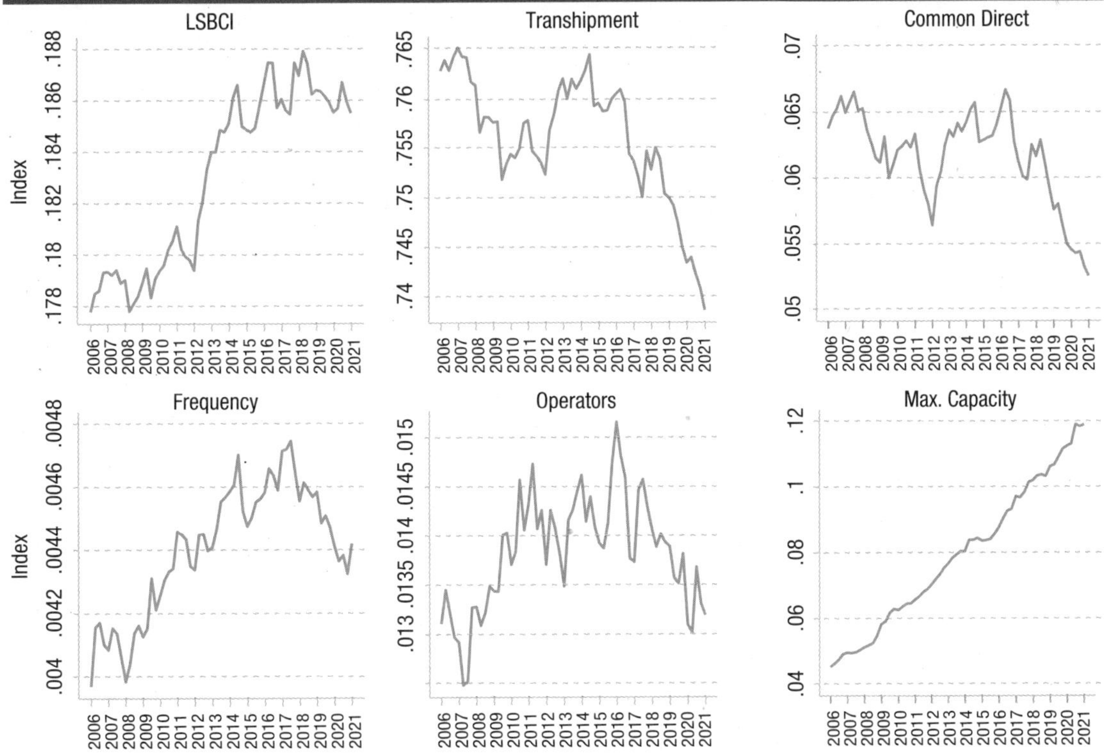

Source: UNCTAD, based on data provided by MDS Transmodal.

All in all, the LSBCI trend reflects a worsening situation for remote and already poorly connected countries. Added to this is the general increase in freight costs which could have severe consequences for international trade (UNCTAD, 2021a).

C. PORT CARGO HANDLING PERFORMANCE

1. Container port performance

In April 2021, to provide stakeholders with a reference point for maritime trade and transport the World Bank and IHS Markit published a new index, the Container Port Performance Index (CPPI) (World Bank 2021, IHS Markit 2021). This index combines data on vessels, their port calls and the cargos they load and unload, as well as the time they spend in ports.

The first version had data for 2019 and the first half of 2020 (table 4.3), and was dominated by ports in East Asia, led by Yokohama in Japan, which was ahead of King Abdullah Port in Saudi Arabia and Qingdao in China. In Europe, the highest-ranked port was Algeciras in Spain at 10; in South Asia, it was Colombo in Sri Lanka at 17; and in the Americas, Lazaro Cardenas in Mexico at 25.

Table 4.3 Top 25 ports under the World Bank IHS Markit Container Port Performance Index 2020

Port name	Economy	Rank
Yokohama	Japan	1
King Abdullah port	Saudi Arabia	2
Chiwan	China	3
Guangzhou	China	4
Kaohsiung	Taiwan Province of China	5
Salalah	Oman	6
Hong Kong	Hong Kong, China	7
Qingdao	China	8
Shekou	China	9
Algeciras	Spain	10
Beirut	Lebanon	11
Shimizu	China	12
Tanjung Pelepas	Malaysia	13
Port Klang	Malaysia	14
Singapore	Singapore	15
Nagoya	Japan	16
Colombo	Sri Lanka	17
Sines	Portugal	18
Kobe	Japan	19
Zhoushan	China	20
Jubail	Saudi Arabia	21
Yosu	Republic of Korea	22
Fuzhou	China	23
Ningbo	China	24
Lazaro Cardenas	Mexico	25

Source: World Bank and IHS Markit Port Performance Program.

The only other North American port in the top 50 was Halifax in Canada. In Africa, the top-ranked port was Djibouti.

UNCTAD has used the raw data from the CPPI to analyse the relationship between the performance of ports and the time ships spend in them. As indicated in figure 4.15 there are clear economies of scale: the more containers there are to load and unload – a larger 'port call size' – the fewer minutes it takes to load or unload a container. Nevertheless, total time in port increases with call size (figure 4.16), so it is reasonable to compare ports or countries within the same range of call sizes.

Port calls where more containers are loaded or unloaded will need longer to handle them, but also be faster for each individual container move, so the correlation between hours in port and speed of handling a slightly negative (figure 4.17). But limiting the analysis to one port call range confirms the expected high positive correlation between the time it takes to move a container and the time it takes to handle a ship (figure 4.18).

For the top 25 economies, table 4.4 summarizes the speed of container handling. For five of the nine call-size ranges the fastest handling is in Taiwan Province of China, followed by Japan for two ranges, and Malaysia and Hong Kong, China for one range each. The ranking per country roughly follows that of the leading individual ports in table 4.3.

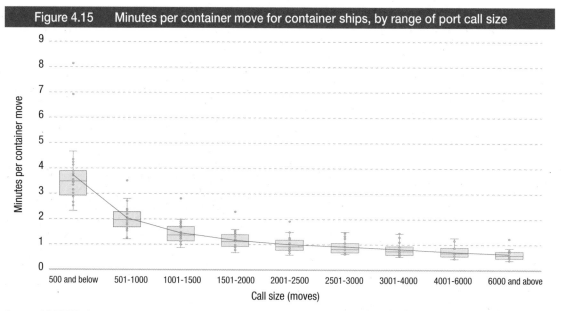

Figure 4.15 Minutes per container move for container ships, by range of port call size

Source: UNCTAD, based on data provided by IHS Markit Port Performance Program.

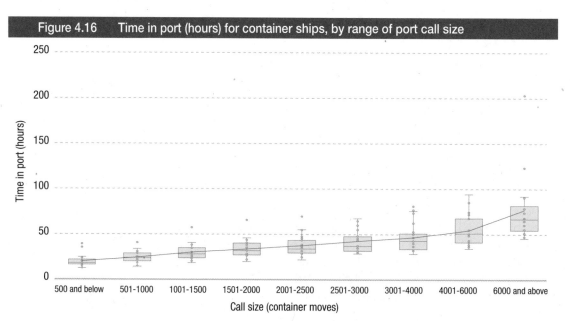

Figure 4.16 Time in port (hours) for container ships, by range of port call size

Source: UNCTAD, based on data provided by IHS Markit Port Performance Program.

Figure 4.17	Correlation between time in port (hours) and minutes per container move, all call sizes

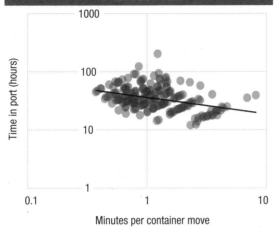

Source: UNCTAD, based on data provided by IHS Markit Port Performance Program.

Figure 4.18	Correlation between time in port (hours) and minutes per container move, only calls with 1001 to 1500 containers per call

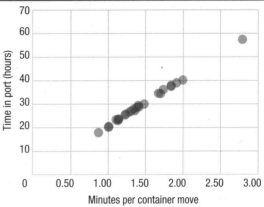

Source: UNCTAD, based on data provided by IHS Markit Port Performance Program. Coefficient of determination (R2) 0.99.

Table 4.4									Minutes per container move, by range of call size, top 25 countries by port calls

Country\call size	<500	501–1000	1001–1500	1501–2000	2001–2500	2501–3000	3001–4000	4001–6000	>6000
Australia	3.44	2.27	1.84	1.57	1.47	1.31	1.28	1.25	0.81
Belgium	3.71	2.08	1.40	1.10	0.91	0.80	0.73	0.70	0.54
Brazil	3.01	1.96	1.48	1.30	1.16	1.07	0.92		
China	2.92	1.68	1.14	0.92	0.77	0.66	0.57	0.49	0.42
Hong Kong, China	3.21	1.60	1.01	0.79	0.77	0.63	0.58	0.45	
Taiwan Province of China	2.31	1.25	0.87	0.67	0.58	0.69	0.51		
France	3.33	2.21	1.70	1.38	1.27	1.23	1.08	0.89	
Germany	4.13	1.92	1.31	1.13	0.96	0.82	0.73	0.65	0.58
India	2.52	1.55	1.22	0.91	0.79	0.75	0.65	0.55	
Indonesia	4.22	2.35	2.00	1.45	1.04	1.00	0.80	0.67	
Italy	3.55	2.41	1.91	1.54	1.46	1.48	1.44	1.14	
Japan	2.57	1.21	1.01	0.80	0.66	0.75	0.70		
Republic of Korea	2.88	1.63	1.14	0.89	0.78	0.75	0.65	0.56	0.70
Malaysia	3.83	2.03	1.38	0.98	0.79	0.69	0.55	0.46	0.37
Netherlands	8.14	2.70	1.67	1.44	1.23	0.99	0.80	0.67	0.62
Panama	4.33	1.86	1.36	1.04	0.94	0.96	0.78	0.88	1.23
Philippines	4.67	3.51	2.79	2.29	1.91	1.43	1.42		
Singapore	3.87	1.81	1.24	0.95	0.76	0.67	0.59	0.47	0.39
Spain	3.87	1.87	1.29	0.98	0.85	0.72	0.63	0.67	0.48
Thailand	2.69	2.79	1.11	0.94	0.79	0.69	0.70	0.66	0.58
Turkey	3.47	2.03	1.42	1.16	1.09	1.06	0.94	0.64	0.57
United Arab Emirates	6.89	2.41	1.74	1.18	0.85	0.70	0.59	0.52	0.41
United Kingdom	3.79	2.18	1.84	1.53	1.28	1.22	1.27	0.93	0.78
United States	3.16	1.77	1.34	1.16	1.06	1.01	0.93	0.90	0.85
Viet Nam	2.64	1.55	1.13	0.78	0.67	0.64	0.58	0.54	0.52
Average	**3.73**	**2.02**	**1.45**	**1.16**	**0.99**	**0.91**	**0.82**	**0.70**	**0.62**
Median	3.47	1.96	1.36	1.10	0.91	0.80	0.73	0.66	0.57
Minimum	2.31	1.21	0.87	0.67	0.58	0.63	0.51	0.45	0.37
Maximum	8.14	3.51	2.79	2.29	1.91	1.48	1.44	1.25	1.23

Source: UNCTAD, based on data provided by IHS Markit Port Performance Program.

Box 4.1 Port performance in Latin America and the Caribbean – differences between types of terminals

In Latin America and the Caribbean across 50 countries and territories, logistics and port services are provided through 1,967 port facilities. Of these, 1,259 are certified as compliant with the International Ship and Port Facility Security (ISPS) Code, including 982 facilities that handle cargo or passenger transfer services, and 277 that provide other services, such as shipyards, docks, and others.

Nonetheless, according to an intensive survey of port facilities in the entire region carried out by the Economic Commission for Latin America and the Caribbean (ECLAC), there are also another 708, of which 590 are port terminals and 118 are related to other types of service.

Port terminals, including those that are ISPS certified and those that are not, represents a widely diverse geographical distribution. The top ten countries according to the number of port terminals are: Brazil, 306; Mexico, 171; Argentina, 143; Chile and Peru, 97 each; Colombia, 88; Paraguay, 65; Bolivarian Republic of Venezuela, 63; Panama, 48 and Cuba, 45. These 10 countries, out of 50, make up 74 per cent of the region's port facilities.

At the opposite end of the ranking, 15 countries or territories have five or fewer facilities each, and almost all have no more than one terminal by port specialty: Antigua and Barbuda, Bermuda, Belize, Barbados, Turks and Caicos Islands, El Salvador, Aruba, Bonaire, Cayman Islands, Dominica, Anguilla, Montserrat, Sint Eustatius, Saint Barth, and Sint Maarten.

A high proportion of these facilities, 470 in total, are multipurpose terminals. The following chart exhibits the distribution by zones and specialties:

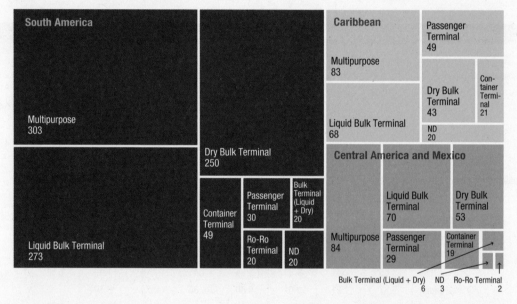

This region is very diverse – in terms of composition, languages, economies, cultural identities, and modes of adaptation to international instruments. The ports systems too differ in terms of maturity and productivity. In the liquid and dry bulk categories, in the most specialized countries, productivity is higher – as in Argentina, Brazil, and Colombia, which move annual volumes close to 600 million mt.

In the last few years region has seen enormous growth in terms of containers, though only four specialized terminals yet have semi-automated processes. Progress in digitalization and paperless transactions has also been slow, and regulatory procedures are not very transparent, making it difficult to promote effective competition. Long-term planning has shown a lack of foresight for ports and connectivity with hinterland infrastructure

Some areas have weakly regulated quasi-monopolistic markets, while others have excessive competition, which may prove harmful. Systems for the design, granting and monitoring of concessions are hampered by institutional weaknesses. These reduce prospects for investment and better multimodal connections and efficient access to markets and ports. The result is often inefficiency and low productivity.

Increasing vertical integration between shipping lines, port terminals and inland logistics heightens the risk of monopoly. In certain areas there are also tensions between management, security, and facilitation. Better security standards would improve development, efficiency, and competitiveness.

Nonetheless, there is some optimism that these problems can be solved – with considerable potential for more containerization and automation of procedures, as well as for improvements in facilitation.

Source: ECLAC, Maritime and Logistics Profile.

2. Dry bulk port performance

VesselsValue[4] has produced a new dataset that combines AIS data on ship movements with data on cargo transfers. This can be used to calculate interesting performance indicators for dry bulk port operations (table 4.5). During the period 2018 to mid-2021, among the top 30 countries in terms of ship arrivals, the average speed of loading ranged from just six ton per minute in Romania and Turkey to 48 ton in Australia.

For dry bulk cargo, unloading tends to be slower than loading, as the operations cannot use the same combination of gravity and conveyer belts. The fastest unloading was in China, at 23 tonnes per minute, and the slowest in Russian Federation, at just 4 tonnes per minute, and in Norway, at just 6 tonnes per minute. These differences partly reflect port performance and economies of scale; Chinese dry bulk terminals are highly mechanized and handle the world's largest iron ore carriers, while Russian Federation and Norway have a long coast with many smaller ports.

Table 4.5 Cargo and vessel handling performance for dry bulk carriers. Top 30 economies by vessel arrivals, average values for 2018 to first half of 2021

	Ton per minute, loading	Ton per minute, discharge	Average waiting to load duration (hours)	Average waiting to discharge duration (hours)
China	19	23	66	56
Australia	48	11	101	50
United States	14	11	101	49
Brazil	25	9	174	131
Russian Federation	12	4	64	71
Canada	17	10	117	70
Argentina	16	7	45	28
South Africa	20	9	83	30
Japan	9	18	43	41
India	14	16	73	63
Ukraine	10	11	55	48
United Arab Emirates	18	10	50	32
Indonesia	10	8	58	54
Republic of Korea	10	16	37	62
New Zealand	10	8	56	26
Chile	11	9	94	94
Turkey	6	9	45	50
Viet Nam	9	11	53	54
Colombia	28	7	39	25
Malaysia	11	13	73	90
Mexico	12	9	68	61
Taiwan Province of China	12	18	34	48
Peru	18	11	82	49
Oman	16	20	80	52
Norway	20	6	84	78
France	10	12	52	55
Saudi Arabia	8	6	49	80
Morocco	8	6	78	127
Romania	6	7	64	29
Mozambique	15	6	94	123

Source: UNCTAD, based on data provided by VesselsValue.
Note: Ranked by number dry bulk carrier arrivals for loading.

---

[4] Data provided electronically by VesselsValue; https://www.vesselsvalue.com, June 2021.

Ships generally wait longer to load than to unload, though there are significant differences between countries. In Colombia, the average waiting time for unloading is one day while in Brazil it is five and a half days. Brazil also has the highest waiting times for loading – on average more than a week. This is partly a consequence of large vessel sizes and longer distances from the main markets. The shortest waits for loading cargo are in Taiwan Province of China at 34 hours. Some countries encourage owners to arrive early to minimize the risk of missing a scheduled port call.

3. Tanker port performance

For tanker port operations too, loading tends to be faster than unloading or 'discharge'. Among the top 30 countries in terms of tanker arrivals, the fastest loading was by the major oil exporters, reaching up to 113 tons per minute for Angola, followed by 95 in Qatar, 90 in Kuwait, and 86 in Saudi Arabia. For unloading oil, the fastest average speeds were in Japan at 83 tons per minute, followed by Republic of Korea at 67 (table 4.6). As regards waiting times, the lowest average time for loading was in Qatar at 26 hours, and for discharge in Japan at 28 hours.

Table 4.6 Cargo and vessel handling performance for tankers. Top 30 countries by vessel arrivals, average values for 2018 to first half of 2021

	Tons per minute, loading	Tons per minute, discharge	Average waiting to load duration (hours)	Average waiting to discharge duration (hours)
United States	24	33	54	69
Russian Federation	38	27	46	36
China	23	43	45	77
Brazil	46	29	62	66
Saudi Arabia	86	31	37	47
United Arab Emirates	66	25	65	89
Republic of Korea	29	67	50	48
Singapore	26	39	47	43
India	26	50	54	68
Malaysia	28	33	47	65
Netherlands	14	29	59	56
Indonesia	19	20	50	62
Italy	15	32	47	48
Mexico	25	17	77	83
Nigeria	43	9	53	129
Kuwait	90	54	32	37
Iraq	50	8	42	96
Canada	37	39	47	62
Spain	15	27	39	37
Qatar	95	48	26	63
Japan	37	83	35	28
United Kingdom	36	26	53	51
Turkey	54	30	36	37
Norway	63	36	46	72
Angola	113	25	37	84
Belgium	12	16	75	42
Bolivarian Republic of Venezuela	20	13	105	79
Taiwan Province of China	22	48	36	40
Argentina	20	20	39	38
Greece	15	30	55	43

Source: UNCTAD, based on data provided by VesselsValue.
Note: Ranked by number tanker arrivals for loading.

E. GREENHOUSE GAS EMISSIONS BY THE WORLD FLEET

1. Shipping is missing its greenhouse gas emissions targets

Over the last decade shipping has become more energy efficient so total emissions have grown slower than the total number of vessels (figure 4.19). Nevertheless, this improvement will not suffice to meet the emissions targets and the agreed objective of the International Maritime Organization (IMO) "to reduce the total annual greenhouse gas emissions by at least 50 per cent by 2050 compared to 2008" as part of the "Initial IMO Strategy on reduction of greenhouse gas emissions from ships" (IMO, 2018).

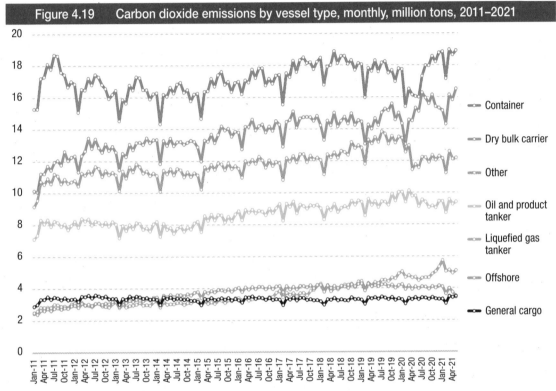

Figure 4.19 Carbon dioxide emissions by vessel type, monthly, million tons, 2011–2021

Source: UNCTAD, based on data provided by Marine Benchmark.

The trends for the world's fleet over the last decade reflect its changing composition, with a declining proportion of journeys for general cargo ships and an increasing one for LNG carriers, with correspondingly higher greenhouse gas (GHG) emissions. In figure 4.19 it is also possible to see the annual downturn in traffic around February in line with the Chinese New Year especially in the dry bulk and container sector.

More recently this chart also shows the impact of the pandemic. 'Other' ships include primarily passenger ships, including ferries and cruise ships which were worst affected. Container ships, also saw an initial decline at the outset of the pandemic but subsequently recovered.

2. Assigning emissions to flag states

Emissions by flag state mostly correspond to market shares for tonnage. But because the fleets have different compositions the ranking is not identical. Liberia, for example, has a larger market share than Marshall Islands in terms of total tonnage (table 2.5), but a far smaller share for CO2 emissions because it has a higher proportion of dry bulk carriers, which produce lower emissions per dwt than other ship types. Germany, on the other hand, is ranked only 29 in the world fleet, but 6 in terms of emissions because a high proportion of its fleet is container ships which tend to go faster than other ship types and emit more CO2 per dwt.[5]

---

[5] Data provided electronically by Marine Benchmark; https://www.marinebenchmark.com, June 2021.

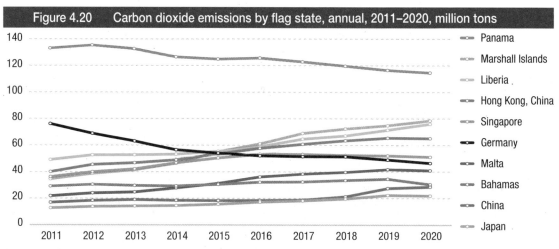

Figure 4.20 Carbon dioxide emissions by flag state, annual, 2011–2020, million tons

Source: UNCTAD, based on data provided by Marine Benchmark.

3. Reducing greenhouse gas emissions may reduce connectivity and increase costs

In June 2021 the IMO's Marine Environment Protection Committee approved a new short-term measure for GHG emissions, with both technical and operational requirements.

Earlier that year, UNCTAD undertook a Comprehensive Impact Assessment of the proposed measure, setting out scenarios for 2030 with or without the measure, across three levels of emission reduction ambition. The aim was to quantify the changes in maritime logistics costs including shipping and time costs. All three indicated an increase in maritime logistics costs.

The IMO subsequently agreed the low scenario, for which the UNCTAD study suggested the following outcomes for 2030:

- A reduction in average speed of 2.8 per cent.
- An increase in average maritime shipping costs by 1.5 per cent.

While significant, these changes are relatively small when compared to typical variations in freight rates. They will also have a very small impact on global GDP and certainly far smaller than the disruption caused by the pandemic or climate change factors, or the costs of not acting in the face of climate change. However the IMO measures will have a greater impact on some countries than others, notably on SIDS or LDCs, which may need support to mitigate the increased costs and alleviate the consequent fallout on their incomes and trade flows (UNCTAD 2021c).

F. SUMMARY AND POLICY CONSIDERATIONS

This chapter has detailed several aspects of port and shipping performance, including fleet deployment and the time ships spend in port, and port performance. It has highlighted persistent differences between ports and countries, and shown how these are shaped by human, institutional, and technological factors.

Developing countries generally perform worse, with higher costs and lower connectivity – a consequence of diseconomies of scale, greater distances from overseas markets, and lower levels of digitalization. These and other countries should be aiming for more competitive commercial environments for port and shipping operations, ensuring that external costs are accounted for.

Costs are likely to increase slightly as a result of measures needed for decarbonization of maritime transport. Smaller and most vulnerable economies may need support to mitigate the increased costs and lower connectivity.

GHG emissions can also be reduced by improving port and shipping performance. If ports can optimize their availability, ships can plan their voyages so as to arrive in port the moment their berth becomes available, thus reducing unnecessary speed and fuel consumption.

Maritime transport will also be transformed by the global energy transition which will increase maritime transport costs and reduce average shipping speeds. Logistics costs increases will be greater for developing than for developed countries.

REFERENCES

IHS Markit (2021). *New Global Container Port Performance Index (CPPI) Launched by the World Bank and IHS Markit*.

IMO (2018). Initial IMO Strategy on Reduction of GHG Emissions from Ships. MEPC 72/17/Add.1 Annex 11. April. Available at https://wwwcdn.imo.org/localresources/en/OurWork/Environment/Documents/ResolutionMEPC.304(72)_E.pdf (accessed 24 May 2020).

MDST (2020). Available at https://www.portlsci.com/index.php (accessed 5 July 2020).

UNCTAD (2017). *Review of Maritime Transport 2017* (United Nations publication. Sales No. E.17.II.D.10. New York and Geneva).

UNCTAD (2021a). Container Shipping in Times of COVID-19: Why Freight Rates Have Surged and Implications for Policy Makers. Policy Brief No. 84. Geneva.

UNCTAD (2021b). Small Island Developing States: Maritime Transport in the Era of a Disruptive Pandemic. Policy Brief, No. 85. UNCTAD. Geneva.

UNCTAD (2021c). UNCTAD Assessment of the Impact of the IMO Short-Term GHG Reduction Measure on States, UNCTAD/DTL/TLB/2021/2, UNCTAD. Geneva.

World Bank (2021). Asian Ports Dominate Global Container Port Performance Index.

This chapter has been prepared in response to a request by the UN General Assembly in its resolution on "International cooperation to address challenges faced by seafarers as a result of the COVID-19 pandemic to support global supply chains"

(A/RES/75/17), at para. 7.

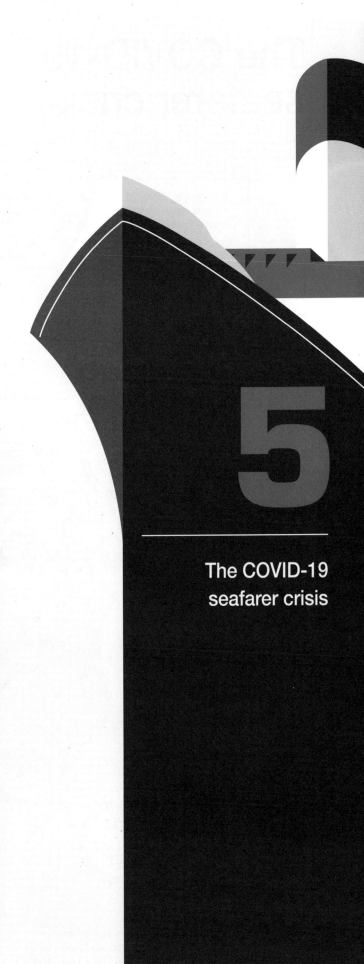

The COVID-19 seafarer crisis

The shipping industry has played a vital role in the global response to the COVID-19 pandemic – delivering food, medical supplies, fuel, and other essential goods, and helping keep global supply chains and commerce running. This is to a large extent due to the world's 1.9 million seafarers, who through these extraordinary times have demonstrated great professionalism and dedication.

But their work has come at some risk to the seafarers themselves, many of whom have been unable to leave their ships. This chapter considers issues related to seafarers' health, safety, security and welfare. It highlights areas where industry, governments, and international organizations can cooperate to protect seafarers' human and legal rights and implement relevant labour standards, including those agreed in the Maritime Labour Convention 2006, and in particular, alleviate their plight resulting from the COVID-19 pandemic. Such support should be part of the implementation of the broader 2030 Agenda – in particular, SDG 8, which aims to ensure decent work and economic growth. Beyond protecting the rights and welfare of seafarers and their families, this would also support the economies of their home countries, help maintain world trade and ensure the flow of goods across supply chains.

Key shipping stakeholders, including international bodies, governments, and industry, have issued guidance and recommendations for ensuring that seafarers are medically fit and have access to medical care, with mechanisms to prevent, and respond to, COVID-19 emergencies at sea – and that ships and port facilities meet international sanitary requirements. They have also argued that seafarers should be recognized as key workers and vaccinated as a matter of priority. However, as the pandemic continues for a second year the crew change crisis appears to be worsening, with continuing logistical obstacles to the repatriation of seafarers. Stakeholders will need to redouble their efforts while also regularly updating their guidance and recommendations in line with the latest scientific insights.

The COVID-19 seafarer crisis

Seafarers, many of whom from developing countries, are playing a vital role in ensuring the flow of critical goods across supply chains and keeping the world trade moving.

All should be working together to implement relevant labour standards, protect seafarers' human rights and advance the objectives of SDG 8 of decent work and economic growth for sustainable development.

Vaccination
Concerted collaborative efforts by industry, governments and international organizations should ensure that seafarers are designated as key workers and are vaccinated as a matter of priority

Crew changes
Governments and industry should continue to work together, including through the Neptune Declaration initiative, and in collaboration with relevant international organizations, to facilitate crew changes, in accordance with international standards and in line with public health considerations

Route deviations
Charterers and other industry stakeholders should be flexible in accepting requests from shipping companies for route deviation to facilitate crew changes

Despite important international efforts and support, the crew change crisis has worsened and seafarers are still facing serious problems which need to be addressed:

1
2
3
4
5
6
7

International legal framework
States and other relevant stakeholders should keep under review the relevant legal framework and ensure that international obligations are respected and implemented

Maritime single windows
Port community systems should implement the Single Window concept to cover all the information and formalities resulting from FAL and other relevant instruments

Information exchange
Relevant public and private sector stakeholders should continue their regular exchange of views and best practices on seafarers' situation and needs

Outbreaks and emergencies at sea
Specific guidance on measures to prevent and deal with COVID-19 and other outbreaks at sea should be updated regularly, in line with developing scientific insights

A. SEAFARERS CRISIS – RECENT DEVELOPMENTS

Shipping and seafarers are vital to global supply chains and the world economy – transporting over 80 per cent of world trade by volume. Around 1.9 million seafarers work to facilitate the way we live, and during the COVID-19 pandemic seafarers have continued to demonstrate great professionalism and dedication, helping to deliver food, medical supplies, fuel, and other essential goods, and keep supply chains active and global commerce running.

Recognizing this, key shipping stakeholders, including international bodies, governments, and industry, have issued guidance and recommendations to support seafarers during the pandemic.[1] The aim is to ensure that seafarers are protected from COVID-19, are medically fit and have access to medical care; that ships and port facilities meet international sanitary requirements; that seafarers are recognised as key workers; and that they are vaccinated as a matter of priority.

However, the pandemic has seriously disrupted crew changes. Each month, a large number of seafarers need to be changed over – to prevent fatigue and to comply with international maritime regulations for safety, crew health and welfare. Aiming to protect public health, as variants of the virus emerge, governments are continuing to impose border closures, lockdowns and preventative measures which include suspending crew changes and prohibiting crews from disembarking at port terminals. Due to these restrictions, and the shortage of international flights, even one year into the pandemic hundreds of thousands of seafarers remain stranded at sea, far beyond the expiration of their contracts (De Beukelaer, 2021). As yet, there is no global consensus on uniform measures that may allow for efficient crew changes and transfer.

The social partners, international organizations, and industry bodies have expressed concern about this humanitarian crisis. IMO, ILO, ICS, ITF, and UNCTAD have urged member States to designate seafarers and other marine personnel as key workers and accept their identity documents as evidence of this status. They have also asked for greater flexibility for ship owners and managers to divert ships and to call in ports where crew change is possible, without imposing penalties. See IMO 2020a, ITF 2020, IMO 2020b, and UNCTAD 2020d.

On 1 December 2020, the UN General Assembly unanimously adopted a resolution on 'International cooperation to address challenges faced by seafarers as a result of the COVID-19 pandemic to support global supply chains' (A/RES/75/17).[2] Indonesia, which supplies much of the maritime labour force, facilitated the negotiation, supported by UNCTAD, ILO and IMO. Co-sponsored by 71 countries, the resolution urges member States to designate seafarers and other marine personnel as key workers and encourages governments and other stakeholders to implement the "Industry recommended framework of protocols for ensuring safe ship crew changes and travel during the Coronavirus (COVID-19) pandemic", the importance of which was recognized by the Maritime Safety Committee of the IMO (IMO, 2021a).[3] The resolution also calls upon governments to facilitate maritime crew changes by enabling them to embark and disembark and expediting travel and repatriation efforts, while also ensuring access to medical care.

In addition, on 8 December 2020, the Governing Body of the International Labour Organization adopted a 'Resolution concerning maritime labour issues and the COVID-19 pandemic' (ILO, 2020b). This urges all Members, to collaborate to identify obstacles to crew changes; designate seafarers as "key workers", for the purpose of facilitation of safe and unhindered movement for embarking or disembarking a vessel, and the facilitation of shore leave. Members should also accept seafarer's internationally recognized documentation, including seafarers' identity documents delivered in conformity with ILO Conventions Nos 108 and 185, and also consider temporary waivers, exemptions or other changes to visa or documentary requirements that might normally apply to seafarers. In addition, they should ensure access

---

[1] See further UNCTAD, 2020a, Chapter 5.E. See also COVID-19-related IMO circulars, https://www.imo.org/en/MediaCentre/HotTopics/Pages/Coronavirus.aspx. For a list of COVID-19 related communications on measures taken by IMO Member states/Associate Members (updated weekly), see http://www.imo.org/en/MediaCentre/HotTopics/Pages/COVID-19-Member-States-Communications.aspx, as well as weekly updates from BIMCO on implementation measures imposed by governments and UN bodies, for sea transport including for crew changes https://www.bimco.org/news/ports/20210528-bimco-covid-19-weekly-report. For calls for action by UNCTAD, see UNCTAD, 2020b, 2020c, 2020d. Also see UNCTAD, 2020e, and 2020f, ILO, 2020, WHO, 2020a, and INTERTANKO, 2020. For a roadmap to improve and ensure good indoor ventilation in the context of COVID-19, see WHO, 2021a. For policy and technical considerations for implementing a risk-based approach to international travel in the context of COVID-19, including for seafarers, see WHO, 2021b and 2021c.

[2] https://undocs.org/en/A/RES/75/17. Inter alia, the Resolution also requests IMO and UNCTAD to report on issues related to the resolution.

[3] Subsequently revised in April 2021, to include reference to vaccination.

to medical facilities ashore, emergency medical treatment and, where necessary, emergency repatriation for seafarers regardless of nationality.

On 21 September 2020, another relevant resolution was adopted by the Maritime Safety Committee of the IMO – 'Recommended action to facilitate ship crew change, access to medical care and seafarer travel during the COVID-19 pandemic' (IMO, 2020c). The IMO urged governments and relevant national authorities to engage nationally and internationally in discussions on the implementation of the industry protocols and consider applying them to the maximum extent possible; designate seafarers as "key workers" providing an essential service, in order to facilitate safe and unhindered movement for embarking or disembarking a vessel; consider temporary measures including (where possible under relevant law) waivers, exemptions or other relaxations from any visa or documentary requirements that might normally apply to seafarers; encourage the use of prevention measures, such as tests on crews before embarkation and provide seafarers with immediate access to medical care ashore.

In response, echoing the above calls, in January 2021, more than 600 companies and organizations signed the 'Neptune Declaration on Seafarer Wellbeing and Crew Change' (Global Maritime Forum, 2021a).[4] The declaration recognizes their shared responsibility to resolve the crew change crisis and calls for the implementation of the industry protocols. For this purpose, it defines four main actions: recognize seafarers as key workers and give them priority access to COVID-19 vaccines; establish and implement gold-standard health protocols based on existing best practice; increase collaboration between ship operators and charterers to facilitate crew changes; and ensure air connectivity between key maritime hubs for seafarers. Subsequently, the signatories developed a set of best practices that serve as a framework for charterers to facilitate crew changes and work with ship owners to minimize the disruptions to operations (Global Maritime Forum, 2021b). In addition, they developed a Neptune Declaration Crew Change Indicator which aggregates data from 10 leading ship managers covering about 90,000 seafarers, to estimate the number affected by the crisis (Global Maritime Forum, 2021c).[5] At the peak of the crisis, more than 400,000 crew were trapped on board their ships. As of March 2021, around 200,000 seafarers remained on board commercial vessels beyond the expiry of their contracts (IMO, 2021b, Aljazeera, 2021).

In March 2021, IMO, ICAO, ILO, WHO and IOM, issued a joint statement on priority vaccination of seafarers and aircrews (IMO, 2021c, IMO, 2021d; ILO, 2021a). Around that time, there were other important documents published, including an industry paper 'COVID-19: Legal, liability and insurance Issues arising from vaccination of seafarers' (ICS et al, 2021a), and a 'Practical guide on vaccination for seafarers and shipowners' (ICS et al, 2021b). A further publication by the ICS in May 2021 was 'Coronavirus (COVID-19): Roadmap for vaccination of international seafarers' (ICS et al, 2021c).

The ILO has a Special Tripartite Committee established under the 2006 Maritime Labour Convention (MLC). In April 2021, the Committee adopted a 'Resolution concerning the implementation and practical application of the MLC, 2006, during the COVID-19 pandemic' which called on Members to designate and treat seafarers as key workers, and take other necessary steps to ensure their rights (ILO, 2021b). This would mean providing them with access to COVID-19 vaccination at the earliest opportunity and promoting the mutual acceptance of vaccine certificates. The Committee also adopted a 'Resolution concerning COVID-19 vaccination for seafarers' (ILO, 2021c), and recommendations concerning the review of maritime-related instruments (ILO, 2021d). In addition, the ILO, following formal requests from shipowner and seafarer organizations, has intervened with member States that have ratified MLC 2006, to remind them of their obligations, notably the obligation of port States to grant access to seafarers in need of medical care in foreign ports (ILO, 2021e).

In May 2021, the IMO Maritime Safety Committee adopted Resolution MSC.490 (103): 'Recommended action to prioritize COVID-19 vaccination of seafarers' (IMO, 2021e), recommending that member States and relevant national authorities prioritize their seafarers, as far as practicable, in their national COVID-19 vaccination programmes, taking into account the WHO SAGE Roadmap (WHO, 2020b). And, while bearing in mind their national vaccines supplies, they should also consider extending COVID-19 vaccines to seafarers of other nationalities.

Seafarers should also be designated as "key workers" and since they frequently travel across borders member States should consider exempting them from requiring proof of COVID-19 vaccination as a condition for entry. In addition, the 109th Session of the International Labour Conference in June 2021

---

[4] Signed by more than 800 companies and organizations, as of June 2021.

[5] Anglo- Eastern, Bernhard Schulte, Columbia Shipmanagement, Fleet Management (FLEET), OSM, Synergy Marine, Thome, V.Group, Wallem, and Wilhelmsen Ship Management.

adopted a 'Global call to action for a human-centred COVID-19 recovery' which prioritizes the creation of decent jobs for all and addresses the inequalities caused by the crisis (ILO, 2021f; ILO, 2021g).

According to IMO, as of the end of June 2021, 60 member States and two associate members had signed on to designate seafarers as key workers (IMO, 2021f). However, despite a gradual easing, many countries still maintain restrictions on crew changes based on nationality or travel history. Problems are also being created in certain contracts of carriage, preventing crew changes while the charterer's cargo is onboard and not allowing the ship to deviate to ports where crew changes could take place (ILO, 2021h; IMO, 2020d). Seafarers also have problems in obtaining visas or travel permits to transit countries.

Despite the above efforts, the crew-change crisis appears to be getting worse. The latest Neptune Declaration Crew Change Indicator published in July 2021 shows that the number of seafarers on board beyond the expiry of their contracts continued to rise in June 2021, as did the number of seafarers on board for over 11 months (table 5.1) (Global Maritime Forum, 2021d). Since the launch of the Indicator in May 2021, the proportion of seafarers on board vessels beyond the expiry of their contract had increased from 5.8 to 8.8 per cent – an increase of over 50 per cent. The number of seafarers on board for over 11 months had increased from 0.4 to 1 per cent – an increase of 150 per cent. According to the MLC 2006, the default maximum period of service on board, following which a seafarer is entitled to repatriation, is 11 months (Regulation 2.5 and Regulation 2.4). In July 2021, the International Chamber of Shipping estimated that, the number of seafarers remaining on board beyond the expiry of their contract, was around 250,000.

Table 5.1   Neptune Declaration Crew Change Indicator, July 2021

	Percentage of seafarers on board beyond the expiry of their contracts		Percentage of seafarers on board for over 11 months	
	Monthly percentage	Percentage point change from previous month	Monthly percentage	Percentage point change from previous month
May 2021	5.8	-	0.8	-
June 2021	7.2	+1.4	0.4	-0.4
July 2021	8.8	+1.6	1.0	+0.6

Source: Global Maritime Forum 2021.

As part of the reporting for the Neptune Declaration Crew Change Indicator, contributing ship managers also highlighted the following key developments: "Continual high infection rates and subsequent domestic lockdowns are still challenging crew changes and causing disruption to crew movements; a decrease of daily inbound flights to the Philippines as well as the travel ban announced by the Philippine Government for seafarers traveling from United Arab Emirates, Oman, Nepal, Bangladesh, Sri Lanka, Pakistan are causing a general disruption to crew movements; travel restrictions continue to prevent seafarers from going back home and many flights have been cancelled; and leading maritime crew nations continue to have low vaccination rates and seafarers continue to have limited vaccine access."[6] (see also box 5.1).

Crew changes and repatriation of seafarers thus still entail serious logistical challenges. Moreover, seafarer access to medical care and priority vaccination remains inadequate, with important repercussions for their health and safety, as well as for public health (DevPolicy, 2021).

In June 2021, it was reported that a cargo ship's captain, who developed COVID-19 symptoms shortly after the vessel set sail, died on board after 11 days (CNN, 2021). Successive ports refused to allow the vessel to call, and no medical evacuation measures were taken. For six weeks, despite repeated pleas for assistance, the ship was stranded offshore, unable to find a port that would take the corpse. As a result, the crew was stuck at sea for weeks, with a potential COVID-19 outbreak on its hands.

This state of affairs is clearly unacceptable. Seafarers should not just be designated as key workers and vaccinated but also provided with speedy and effective emergency medical assistance in the event of a COVID-19 outbreak at sea.

It will also be important to keep abreast of the latest guidance, which should be updated in line with the latest scientific insights on transmission pathways, variants, vaccine efficacy, and related risks.

---

[6] According to ICS, informal industry survey data about vaccinations by nationality of seafarers suggests that, with some notable exceptions, only a small proportion of the world's seafarers has been currently vaccinated.

The latest industry guidance for ship-operators (ICS et al., 2021d), draws on sector-specific WHO guidance published in August 2020 (WHO, 2020a).[7] A good model is that of Belgium which in July 2021, started a vaccination programme for all seafarers arriving in a Belgian port, regardless of their nationality (Safety4Sea, 2021). Other countries have seafarer vaccination programmes, including Australia, Cyprus, Germany, the Netherlands, and the United States. In India the National Union of Seafarers has started a programme to offer 5,000 doses to seafarers and their families (TradeWinds, 2021).

Addressing the complex issues arising in the context of facilitating global trade in times of a pandemic while protecting the health of seafarers and the public at large will require the continued engagement of all stakeholders, including in the negotiations of legal instruments, guidelines and recommendations under the auspices of UN bodies, including ILO, IMO, and UNCTAD, and in respect of relevant national and local implementation. Reflecting the continued need to raise awareness and alleviate the plight of seafarers, while recognizing their vital role in world trade, it is worth noting that "Seafarers: at the core of shipping's future" was selected as the World Maritime theme for 2021.[8]

According to the BIMCO/ICS Seafarer Workforce Report 2021 (BIMCO/ICS 2021), in 2021 around the world there were 1,892,720 seafarers, of whom 857,540 were officers and 1,035,180 were ratings – skilled seafarers who carry out support work for officers. The largest supplier for both officers and ratings was the Philippines followed by the Russian Federation, Indonesia, China, and India (table 5.2). Together, these countries supplied 44 per cent of the global seafarer workforce. These numbers are growing.

Box 5.1　　The case of the Philippines

Seafarer supply

The Philippines is now the world's largest source of seafarers, with an estimated 700,000 deployed on domestic or foreign-flagged seagoing vessels. Over a quarter of all global merchant shipping crew members come from the Philippines. As of 2019, there were 380,000 Filipino seafarers overseas. By mid-2020, over the three months after the onset of the COVID-19 pandemic and the quarantine imposed in the country, 50,000 Filipino seafarers had been repatriated, but only 17,845 outbound or deployed seafarers were recorded by the authorities. As reported by Business Mirror, during July–September 2020, according to the Philippine Overseas Employment Administration, the deployment of Filipino seafarers started to return to normal, with over 136,000 sailors able to board ships traveling in international waters.

Seafarer remittances

In 2019, the Philippines earned more than $30.1 billion from overseas Filipino workers, including $6.5 billion from seafarers. In 2019, the remittances of overseas Filipino workers constituted 9.3 per cent of the Philippines' GDP and 7.3 per cent of gross national income. By the end of 2020, total remittances of overseas foreign workers amounted to $29.9 billion a 0.8 per cent decline that year. Of this amount, $6.3 billion was remitted by sea-based workers – a 2.8 per cent decline.

Seafarer vaccination

When it comes to vaccination against COVID-19, seafarer-supplying nations are at a disadvantage. According to the *New York Times* vaccination tracker, as of the beginning of August 2021, globally on average 53 doses of the COVID-19 vaccines had been administered for every 100 people, but the Philippines had delivered only 18 doses for every 100 people. Among the world's five-largest seafarer providers every country except China (117) had delivered less than the global average: Russian Federation, 42; Indonesia, 25; and India, 34.

Sources: Maritime Industry Authority (2020). A Letter to All Filipino Seafarers Around the World.13 April. https://marina.gov.ph/2020/04/13/a-letter-to-all-filipino-seafarers-around-the-world. The World Bank. https://data.worldbank.org. Philippines Overseas Employment Administration https://www.poea.gov.ph/ofwstat/ofwstat.html. Global Maritime Forum (2020). S.E.A.F.A.R.E.R. 30 September. https://www.globalmaritimeforum.org/news/s-e-a-f-a-r-e-r. Business Mirror (2020). 136,000 Filipino seafarers deployed aboard international vessels overseas since July'. 1 October. https://businessmirror.com.ph/2020/10/01/136000-filipino-seafarers-deployed-aboard-international-vessels-overseas-since-july. Bangko Sentral ng Pilipinas (2021). Statistics. Overseas Filipinos' Cash Remittances. https://www.bsp.gov.ph/statistics/external/ofw2.aspx. Philippine Statistics Authority. https://psa.gov.ph/national-accounts/base-2018/data-series.

---

[7]　The industry guidance also refers to non-sector specific guidance for the general public (WHO, 2020c).

[8]　https://www.imo.org/en/MediaCentre/PressBriefings/pages/DOTS-2021.aspx.

Table 5.2 Five largest seafarer-supply countries, 2021

	All Seafarers	Officers	Ratings
1	Philippines	Philippines	Philippines
2	Russian Federation	Russian Federation	Russian Federation
3	Indonesia	China	Indonesia
4	China	India	China
5	India	Indonesia	India

Source: BIMCO/ICS, Seafarer Workforce Report 2021, London, 2021.

The 1978 International Convention on Standards of Training, Certification and Watchkeeping for Seafarers (STCW) establishes basic requirements on training, certification and watchkeeping. Between 2015 and 2021 the supply of STCW-certified officers increased by 11 per cent and that of STCW-certified ratings by 19 per cent (BIMCO/ICS 2015).

B. SEAFARER CRISIS – IMPLEMENTATION OF THE ILO MARITIME LABOUR CONVENTION, 2006, AS AMENDED (MLC 2006)

The ILO Maritime Labour Convention 2006, entered into force on 20 August 2013 and, as of July 2021, had been ratified by 98 of the 187 ILO member States. The Convention comprehensively sets out rights and responsibilities, as well as minimum standards for seafarers' working and living conditions. It covers a wide range of issues, including minimum age, employment agreements, hours of work or rest, payment of wages, paid annual leave, repatriation at the end of contract, and onboard medical care. It also addresses licensed private recruitment and placement services, accommodation, food and catering, health and safety protection and accident prevention and complaint handling. In addition, the Convention introduces compliance and enforcement components for flag State inspection and for port State control. The MLC 2006,[9] taken together with other instruments, thus helps guarantee the health, safety, security and welfare of seafarers as well as their human rights.[10]

Nevertheless, as result of COVID-19 restrictions many seafarers have been stranded. As a recent UN report highlights, "hundreds of thousands of seafarers are trapped on ships as routine crew changes cannot be carried out, while hundreds of thousands are stranded on land, prevented from re-joining ships. Those stranded on ships are being denied their human rights, including their rights to physical and mental health, to family life, and to freedom of movement, and are often forced to work beyond the default 11-month maximum period of service on board, as established by MLC 2006. This is resulting in cases that could amount to forced labour" (UN Global Compact, et al., 2021). The report addresses seafarers' rights, and offers cargo owners, charterers and logistics providers guidance and a checklist for conducting due diligence across their supply chains. The aim is to identify, prevent, mitigate and address adverse human rights impacts for seafarers affected by the ongoing COVID-19 crisis.

On 12 December 2020, the ILO Committee of Experts on the Application of Conventions and Recommendations, adopted a document entitled 'General observation on matters arising from the application of the MLC, 2006, during the COVID-19 pandemic' (ILO, 2021i). The Committee noted with deep concern the impact that COVID-19 restrictions have had on the protection of seafarers' rights as laid out in the Convention. The Committee also took note of the observations of the International Transport Workers' Federation received on 1 October 2020 and of the International Chamber of Shipping on 26 October 2020 that ratifying States had failed to comply with major provisions of the Convention during the COVID-19 pandemic – notably regarding cooperation among Members, access to medical care and repatriation of seafarers. In addition, they noted the risk that fatigue and other health issues could lead to serious maritime accidents.[11] It therefore, strongly encouraged ratifying States in their different capacities as flag States, port States or labour-supplying States that have not yet done so, "to recognize seafarers as key workers without delay and to draw in practice the consequences of such qualification, in order to restore the respect of their rights as provided for in the MLC, 2006."

---

[9] https://www.ilo.org/dyn/normlex/en/f?p=NORMLEXPUB:91:0::NO::P91_INSTRUMENT_ID:312331.

[10] The protection of human rights is a cross cutting issue for the 2030 Agenda for Sustainable Development, which seeks to realize the human rights of all (see A/RES/70/1, Preamble). Thus, the 2030 Agenda and human rights are interwoven and inextricably tied together (OHCHR, 2015).

[11] For further information on the labour rights and standards involved, see ILO, 2020c, 2020d.

In February 2021, ILO, through a revised information note, published guidance, on how best to address the complexities of the current crisis in light of the provisions of MLC, 2006. This was updated to reflect the observations of the ILO Committee of Experts on the Application of Conventions and recommendations (ILO, 2021j), and also made reference to the MLC, 2006 and previous work of ILO bodies[12], as well as to recommendations from the IMO and WHO, and related work by the ICS and the ITF.

The Committee advises that the notion of 'force majeure', i.e., unforeseen or unforeseeable circumstances making it impossible to comply with the MLC 2006, may no longer be invoked from the moment that options are available to comply with the provisions of the Convention, although more difficult or cumbersome, and urged ratifying States which have not yet done so, to adopt all necessary measures without delay to restore the protection of seafarers' rights and comply to the fullest extent with their obligations under the MLC 2006.

The note urges all *ratifying States* to:

- Adopt the necessary measures or reinforce existing ones without delay to ensure that, in no case, are seafarers forced to continue working on extended contractual arrangements without their formal, free, and informed consent.

- Recognize seafarers as key workers without delay and to draw in practice the consequences of such qualification, in order to restore the respect of their rights as provided for in the MLC, 2006.

- Adopt necessary measures, in consultation with relevant seafarers' and shipowners' organizations, to further enhance cooperation with each other to ensure the effective implementation and enforcement of the Convention, in particular during the COVID-19 pandemic.

Flag States are urged to ensure that:

- The ships that fly their flags fully comply with the provisions of the Convention and adopt the necessary measures and/or reinforce the existing ones without delay, including through more frequent inspections, if necessary.

- Seafarers on ships that fly their flags are covered by adequate measures for the protection of their health and have access to prompt and adequate medical care whilst working on board, including access to vaccination (Regulation 4.1).

- Seafarers are provided with occupational health protection and live, work and train on board ship in a safe and hygienic environment (Regulation 4.3).

- The prohibition to forgo minimum annual leave with pay is strictly enforced, with the limited exceptions authorized by the competent authority (Regulation 2.4 and Standard A2.4, paragraph 3).

- Seafarers are repatriated at no cost to themselves in the circumstances specified in the Convention, with strict respect of the default 11 months maximum period of service on board derived from the provisions of the Convention (Regulation 2.5 and Regulation 2.4).

- Ships that fly their flag have sufficient of seafarers employed on board to ensure that ships are operated safely, efficiently and with due regard to security under all conditions, taking into account concerns about seafarer fatigue and the particular nature and conditions of the voyage (Regulation 2.7).

- No fees or other charges for seafarer recruitment or placement, including the cost of any quarantine obligations before joining the ship, are borne directly or indirectly, in whole or in part, by the seafarer, other than the cost authorized under Standard A1.4, paragraph 5.

- Seafarers are granted shore leave for their health and well-being and consistent with the operational requirement of their positions, subject to the strict respect of any public health measures applicable to the local population.

Port States are urged to:

- Ensure that seafarers on board ships in their territory who are in need of immediate medical care, are given access to medical facilities on shore (Regulation 4.1).

- Facilitate the repatriation of seafarers serving on ships which call at their ports or pass through their territorial or internal waters (Standard A2.5.1, paragraph 7).

---

[12] Including the CEACR and the Special Tripartite Committee of MLC 2006.

- Allow and facilitate the replacement of seafarers who have disembarked and consequently ensure the safe manning of ships, by providing an expeditious and non-discriminatory treatment of new crew members who enter their territory exclusively to join their ships (Standard A2.5.1, paragraph 7).

Labour-supplying States which have not yet done so, are called upon to:

- adopt the necessary and immediate measures to ensure that the required facilities are put in place in relation to transport, testing and quarantine of seafarers.

- While encouraging a pragmatic approach regarding certificates in respect of training and qualifications since the beginning of the pandemic, *all ratifying States* are urgently called upon to adopt all necessary measures without delay to restore the protection of seafarers' rights and comply, to the fullest extent, with their obligations under the MLC 2006.

- With respect to maritime labour certificates and inspections, while recognizing challenges since the outbreak of COVID-19, in respect of conducting the inspections required in accordance with MLC 2006, all ratifying countries with responsibilities as *flag States* and *port States* are urged to adopt the necessary measures without delay, to ensure compliance with the Convention.

In addition, the guidance notes that the measures adopted to contain the pandemic are creating additional challenges in resolving the cases of abandonment that occurred before the outbreak of COVID-19. The IMO/ILO database on reported incidents of abandonment of seafarers, shows a dramatic increase in cases of abandonment in the second part of 2020, with some of those cases linked to COVID-19-related measures.[13] It was recalled that, even in the context of the COVID-19 pandemic, flag States, port States and labour-supplying States remain bound by the requirements concerning repatriation set out in Regulation 2.5 of the MLC 2006, and the relevant provisions of the Code of the Convention.

Member States must undertake all necessary action to promptly resolve situations of abandonment and ensure that affected seafarers are repatriated as soon as possible and receive the payment of outstanding wages, in accordance with the relevant provisions of the MLC 2006 (ILO, 2021j). According to ILO, as of mid-July 2021, 60 cases had been reported for 2021, which, if that rate continued, would surpass the number of cases in 2020. Also, resolution of a number of abandonment cases had been delayed due to the pandemic (e.g., not being able to repatriate seafarers due to restrictions on disembarkation and travel).

C. CREW CHANGES AND KEY WORKER STATUS – OTHER RELEVANT INTERNATIONAL LEGAL INSTRUMENTS

In addition to the MLC 2006, a number of other international conventions and instruments contain provisions aiming to reduce the formalities and documents required, and to facilitate and simplify crew changes. These cover issues such as seafarers' repatriation, transit and joining ships, and the issuance and harmonization of seafarers' identity documents, while enhancing border and port security. Adopting and implementing these instruments would ease the situation of seafarers during the COVID-19 pandemic and beyond.

ILO Convention No. 108 on Seafarers' Identity Documents, 1958

It has been a longstanding common practice to allow seafarers shore leave to access medical, communications and other onshore welfare facilities. In addition, to join or change ships seafarers may need to transit or transfer through countries, which requires border facilitation at seaports and airports. For this purpose, they have traditionally been issued with a seafarers' identity document (SID). Although a SID is not considered a travel documents per se, like a passport or visa it may be subject to the same national laws.

The Seafarers' Identity Documents Convention, 1958 (No. 108) entered into force on 19 February 1961, and has been ratified by 64 States.[14] The Convention specifies the minimum mandatory details that should be contained in the SID but does not require any security features, or specific form of the document. As a result, various countries subsequently developed their own, making it difficult for border and port authorities to determine whether a document is legitimate.

---

[13] https://www.ilo.org/dyn/seafarers/seafarersbrowse.home. For more information on work by IMO/ILO in cooperation with ITF, on the issue of abandonment of seafarers, see https://www.imo.org/en/OurWork/Legal/Pages/Seafarer-abandonment.aspx.

[14] https://www.ilo.org/dyn/normlex/en/f?p=NORMLEXPUB:12100:0::NO::P12100_ILO_CODE:C108.

ILO Convention No. 185 on Seafarers' Identity Documents (Revised) 2003, as amended

Following the terrorist attacks of 11 September 2001, the Seafarers' Identity Documents Convention (Revised), 2003 (No. 185), was adopted.[15] It included innovations that related to the form of the SID, which addition to a photograph and other details could include biometric security features such as fingerprints as well as verification options for uniformity and machine readability. The Convention also contains minimum requirements for the SID's issuance processes and procedures, including quality control, national databases, and national focal points to provide information to border authorities. In particular, article 6, paragraph 7, of the Convention, provides: "Each Member for which this Convention is in force shall, in the shortest possible time, also permit the entry into its territory of seafarers holding a valid seafarers' identity document supplemented by a passport, when entry is requested for the purpose of: (a) joining their ship or transferring to another ship; (b) passing in transit to join their ship in another country or for repatriation; or any other purpose approved by the authorities of the Member concerned."

Convention No.185 entered into force in February 2005, but so far has been ratified by only 36 out of 187 ILO member States, including only few port States. Although some countries have made considerable investment to properly implement this Convention, they can therefore only count on only a few other countries to recognize their SIDs. Moreover, only a few ratifying countries are in a position to issue SIDs that conform with the Convention, while 64 countries still remain Parties only to the 1958 Convention.

Implementation has been slow partly because the specified fingerprint technology and biometric features were soon considered out of date. Instead, since 2003 many border authorities have been using the standards of the International Civil Aviation Authority, namely, ICAO Doc 9303 on Machine Readable Travel Documents.[16] This is now universally followed for travel and similar documents and includes the facial image in a contactless chip – as in electronic passports.

In 2016, ILO Convention No.185 was subsequently amended to align its biometric requirements with those of ICAO Doc 9303.[17] This way, the SID should look and function like an e-passport, booklet, or card and can be issued, read, and verified with the same equipment – enhancing security while simplifying the processes for seafarers when they arrive in ports, or transit or cross international borders.

The amended version entered into force in June 2017, and the amendments are applicable to all member States to the original Convention No.185, except for Marshall Islands. Authorities issuing SIDs, were given a five-year transition period to update their systems, i.e., until 2022, although individual countries may issue the new SIDs as soon as they are able to. All the 1.9 million seafarers could benefit from the new SIDs, which would allow them to travel without a visa to join their ships and to disembark in ports. Unfortunately, implementation appears to have slowed due to the COVID-19 pandemic.

IMO Convention on Facilitation of International Maritime Traffic, 1965 (FAL Convention)

The IMO Convention on Facilitation of International Maritime Traffic, 1965 (FAL Convention) entered into force on 5 March 1967, and has been ratified by 125 out of 174 IMO member States.[18] Its objective is "to facilitate maritime traffic by simplifying and reducing to a minimum the formalities, documentary requirements and procedures on the arrival, stay and departure of ships engaged in international voyages." Rather than address trade-related aspects of shipping, it focuses on the formalities and procedures for ships calling in ports, including those related to the arrival and departure of seafarers.

The FAL Convention contains standards and recommended practices and rules for simplifying formalities and documentary requirements. Customs and immigration officials and port authorities should ask for the minimum of information at the appropriate time, and offer documents to be completed in a standard format, while those providing information, should provide accurate data, at the appropriate time and in the agreed format.

---

[15] https://seafarersrights.org/wp-content/uploads/2018/03/INTERNATIONAL_TREATY_ILO-CONVENTION-C185_2003_ENG.pdf.

[16] https://www.icao.int/publications/pages/publication.aspx?docnum=9303.

[17] https://www.ilo.org/dyn/normlex/en/f?p=1000:12100:0::NO::P12100_ILO_CODE:C185.

[18] https://euroflag.lu/wp-content/uploads/2019/03/Convention-on-Facilitation-of-International-Maritime-Traffic-1965-as-amended-FAL-Convention-2.7.4-Recommended-Practice.pdf. For more information on the FAL Convention, see Chapter 5, *Review of Maritime Transport 2021*.

- *2009 amendments to the FAL Convention*[19] – These entered into force on 15 May 2010 and include changes related to the contents and purpose of documents: "A passport or an identity document issued in accordance with relevant ILO conventions, or else a valid and duly recognized seafarer's identity document, shall be the basic document providing public authorities with information relating to the individual member of the crew on arrival or departure of a ship."

- *2016 amendments to the FAL Convention*[20] – These entered into force on 1 January 2018 and provide for additional guarantees. Any discrimination is prohibited, and shore leave should be granted to crew members, irrespective of the ship's flag State. Since 2019, ships and ports have had to exchange FAL data electronically and are encouraged to use a "single window", in which all the many agencies and authorities exchange data via a single point of contact. Following the expected adoption of further amendments in 2022, and their subsequent entry into force, the single window could become obligatory from January 2024.

The IMO Compendium on Facilitation and Electronic Business[21]

This is an important IMO instrument for accelerating digitalization and connectivity in the maritime industry. It facilitates the exchange of information ship to shore and enables interoperable single windows – reducing port formalities by harmonizing the data elements required and standardizing electronic messages. Its key components are the IMO Data Set and the IMO Reference Data Model which provide common semantics and representation of the data needed to fulfil ship reporting requirements. The IMO data elements are mapped across the main models (e.g., UN/CEFACT, WCO Data Model and ISO) ensuring full interoperability between standards for ship clearance. Since 2019, the Compendium has been extended beyond FAL forms and is now connected to several IMO instruments, such as MARPOL for advance notification of waste delivery to port reception facilities. From 2020, the Compendium also included the Maritime Declaration of Health (MDH), a requirement of the WHO International Health Regulations.

IMO Guidelines for setting up a maritime single window

The IMO has developed guidelines for setting up a maritime single window (MSW).[22] These offer information, advice and guidance along with examples of the experience and knowledge gained by some member States in introducing an MSW. Single windows, mainly for cargo, are currently being developed under various technical assistance projects in developing countries, including in cooperation with ASYCUDA.[23] MSW and port community systems can smooth formalities, (e.g., data elements included in the crew list, the passenger list and the maritime declaration of health).[24]

D. THE WAY FORWARD

Despite important international support, seafarers are still facing serious problems as a result of the COVID-19 pandemic. This requires urgent action in a number of important areas.

- *Vaccination* – Concerted collaborative efforts by industry, governments and international organizations should ensure that seafarers are designated as key workers and are vaccinated as a matter of priority.

- *Crew changes* – Governments and industry should continue to work together, including through the Neptune Declaration initiative, and in collaboration with relevant international organizations, to facilitate crew changes, in accordance with international standards and in line with public health considerations. They should also ensure the availability and access to related seafarer data.

- *Route deviations* – Charterers and other industry stakeholders should be flexible in accepting requests from shipping companies for route deviation to facilitate crew changes and should refrain from using "no crew change" clauses in charterparties.

---

[19] https://www.imo.org/fr/MediaCentre/MeetingSummaries/Pages/FAL-35th-Session.aspx.

[20] https://www.imo.org/fr/MediaCentre/MeetingSummaries/Pages/FAL-40th-session.aspx.

[21] https://www.imo.org/en/OurWork/Facilitation/Pages/IMOCompendium.aspx.

[22] https://wwwcdn.imo.org/localresources/en/OurWork/Facilitation/Facilitation/FAL.5-CIRC.42-REV.1.pdf.

[23] https://asycuda.org/en/. Also see Chapter 6, part on trade facilitation.

[24] For further information on Single Windows, see Chapter 5 of the *Review of Maritime Transport 2021*. Also see Premti A., Asariotis R., 2021.

- *International legal framework* – States and other relevant stakeholders should, in consultations and meetings on seafarers' issues at ILO and IMO, keep under review the relevant legal framework and ensure that international obligations are respected and implemented.
- *Maritime single windows* – Port community systems should implement the Single Window concept, similarly to the customs-centric Single Window powered by ASYCUDA, to cover all the information and formalities resulting from FAL and other relevant instruments.
- *Information exchange* – Relevant public and private sector stakeholders should continue their regular exchange of views and best practices on seafarers' situation and needs, and lessons learned, including from the COVID-19 pandemic, and promote further harmonization and standardization.
- *Outbreaks and emergencies at sea* – In line with developing scientific insights, governments, international organizations and all stakeholders should regularly update specific guidance on measures to prevent and deal with COVID-19 and other outbreaks at sea and ensure that mechanisms are in place to reduce, and respond to medical emergencies at sea.

Public and private stakeholders must continue to work together to implement relevant labour standards and address health, safety, security, welfare, and other challenges faced by seafarers. All should be working to protect seafarers' human rights and advance the objectives of SDG 8 of decent work and economic growth for sustainable development.

REFERENCES

Aljazeera (2021). Abandoned: The seafarers stuck at sea for two years. 6 July. Available at https://www.aljazeera.com/features/2021/7/6/abandoned-the-seafarers-stuck-onboard-for-two-years.

Bangko Sentral ng Pilipinas (2021). Statistics. Overseas Filipinos' Cash Remittances. Available at https://www.bsp.gov.ph/statistics/external/ofw2.aspx.

BIMCO/ICS (2015). Manpower Report: The Global Supply and Demand for Seafarers in 2015. Available at https://www.ics-shipping.org/wp-content/uploads/2020/08/manpower-report-2015-executive-summary.pdf.

BIMCO/ICS (2021). Seafarer Workforce report. The Global Supply and Demand for Seafarers in 2021. Available at https://www.ics-shipping.org/publication/seafarer-workforce-report-2021-edition/.

Business Mirror (2020). 136,000 Filipino seafarers deployed aboard international vessels overseas since July. 1 October. Available at https://businessmirror.com.ph/2020/10/01/136000-filipino-seafarers-deployed-aboard-international-vessels-overseas-since-july.

CNN (2021). This cargo ship's captain died aboard. Then the crew was stuck at sea for weeks – CNN. 21 June. Available at https://edition.cnn.com/2021/06/19/europe/italy-captain-capurro-intl-hnk-dst/index.html.

De Beukelaer (2021). COVID-19 border closures cause humanitarian crew change crisis at sea. Marine Policy. June. Available at https://www.sciencedirect.com/science/article/pii/S0308597X21002724?via%3Dihub.

DevPolicy, 2021. Stranded seafarers: an unfolding humanitarian crisis. 3 June. Available at https://devpolicy.org/seafarers-in-a-covid-world-20210603/.

Global Maritime Forum (2020). S.E.A.F.A.R.E.R. 30 September. Available at https://www.globalmaritimeforum.org/news/s-e-a-f-a-r-e-r.

Global Maritime Forum (2021a). The Neptune Declaration on Seafarer Wellbeing and Crew Change. Available at https://www.globalmaritimeforum.org/content/2020/12/The-Neptune-Declaration-on-Seafarer-Wellbeing-and-Crew-Change.pdf.

Global Maritime Forum (2021b). Best Practices for Charterers. Available at https://www.google.com/url?sa=t&rct=j&q=&esrc=s&source=web&cd=&ved=2ahUKEwj80t-Rpf_wAhXqAWMBHQOiBREQFnoECAgQAA&url=https%3A%2F%2Fwww.globalmaritimeforum.org%2Fcontent%2F2021%2F05%2FThe-Neptune-Declaration-Best-Practices-for-Charterers.pdf&usg=AOvVaw3DEFTUi6CcR8t570J4DvEn.

Global Maritime Forum (2021c). The Neptune Declaration Crew Change Indicator. Available at https://www.globalmaritimeforum.org/content/2021/05/The-Neptune-Declaration-Crew-Change-Indicator-Data-collection-and-processing.pdf.

Global Maritime Forum (2021d). The Neptune Declaration Crew Change Indicator. July 2021. Available at https://www.globalmaritimeforum.org/content/2021/06/The-Neptune-Declaration-Crew-Change-Indicator-July-2021.pdf.

ICS et al (2021a). Coronavirus (COVID-19): Legal, Liability and Insurance Issues arising from Vaccination of Seafarers. March. Available at https://www.ics-shipping.org/publication/coronavirus-covid-19-legal-liability-and-insurance-issues-arising-from-vaccination-of-seafarers/.

ICS et al (2021b). Coronavirus (COVID-19) Vaccination for Seafarers and Shipping Companies: A Practical Guide. March. Available at https://www.ics-shipping.org/publication/coronavirus-covid-19-vaccination-practical-guide/.

ICS et al (2021c). Coronavirus (COVID-19): Roadmap for Vaccination of International Seafarers. May. Available at https://www.ics-shipping.org/publication/coronavirus-covid-19-roadmap-for-vaccination-of-international-seafarers/.

ICS et al (2021d). Coronavirus (COVID-19): Guidance for Ship Operators for the Protection of the Health of Seafarers, Fourth Edition. 7 June. Available at https://www.ics-shipping.org/publication/coronavirus-covid-19-guidance-fourth-edition/.

ILO (2020a). COVID-19 and maritime shipping & fishing. Briefing note. 15 May. Available at https://www.ilo.org/sector/Resources/publications/WCMS_742026/lang--en/index.htm.

ILO (2020b). Resolution concerning maritime labour issues and the COVID-19 pandemic. 8 December. Available at https://www.ilo.org/wcmsp5/groups/public/---ed_norm/---relconf/documents/meetingdocument/wcms_760649.pdf.

ILO (2020c). COVID-19 and the world of work: Impact and policy responses 18 March. Available at https://www.ilo.org/wcmsp5/groups/public/---dgreports/---dcomm/documents/briefingnote/wcms_738753.pdf.

ILO (2020d). New Statement of the Officers of the STC 1 on the coronavirus disease (COVID-19). 1 October. Available at https://www.ilo.org/wcmsp5/groups/public/---ed_norm/---normes/documents/statement/wcms_756782.pdf.

ILO (2021a). Seafarers and aircrew need priority COVID-19 vaccination. 26 March. Available at https://www.ilo.org/global/about-the-ilo/newsroom/news/WCMS_776797/lang--en/index.htm.

ILO (2021b). Resolution concerning the implementation and practical application of the MLC, 2006 during the COVID-19 pandemic. Available at https://www.ilo.org/wcmsp5/groups/public/@ed_norm/@normes/documents/genericdocument/wcms_782881.pdf.

ILO (2021c). Resolution concerning COVID-19 vaccination for seafarers. Available at https://www.ilo.org/wcmsp5/groups/public/@ed_norm/@normes/documents/genericdocument/wcms_782880.pdf.

ILO (2021d). Recommendations concerning the review of maritime-related instruments. Available at https://www.ilo.org/wcmsp5/groups/public/@ed_norm/@normes/documents/genericdocument/wcms_783227.pdf.

ILO (2021e). Background paper for discussion. Fourth meeting of the Special Tripartite Committee established under Article XIII of the Maritime Labour Convention, 2006, as amended – Part I (Geneva, 19–23 April 2021). Available at https://www.ilo.org/wcmsp5/groups/public/---ed_norm/---normes/documents/genericdocument/wcms_774787.pdf.

ILO (2021f). Resolution concerning a global call to action for a human-centred recovery from the COVID-19 crisis that is inclusive, sustainable and resilient. 17 June. Available at https://www.ilo.org/ilc/ILCSessions/109/reports/texts-adopted/WCMS_806092/lang--en/index.htm.

ILO (2021g). Work in the time of COVID Report of the Director-General International Labour Conference 109th Session, 2021. Available at https://www.ilo.org/wcmsp5/groups/public/---ed_norm/---relconf/documents/meetingdocument/wcms_793265.pdf.

ILO (2021h). Statement of the Officers of the STC on the coronavirus disease (COVID-19) regarding increased collaboration between shipowners and charterers to facilitate crew changes. 15 December. Available at https://www.ilo.org/global/standards/maritime-labour-convention/special-tripartite-committee/WCMS_764724/lang--en/index.htm.

ILO (2021i). General observation on matters arising from the application of the Maritime Labour Convention, 2006, as amended (MLC, 2006) during the COVID-19 pandemic. Available at https://www.ilo.org/wcmsp5/groups/public/---ed_norm/---normes/documents/publication/wcms_764384.pdf.

ILO (2021j). Information note on maritime labour issues and coronavirus (COVID-19). Revised version 3.0. 3 February. Available at https://www.ilo.org/wcmsp5/groups/public/---ed_norm/---normes/documents/genericdocument/wcms_741024.pdf.

IMO (2020a). Coronavirus (COVID-19) – Joint Statement calling on all Governments to immediately recognize seafarers as key workers, and to take swift and effective action to eliminate obstacles to crew changes, so as to address the humanitarian crisis faced by the shipping sector, ensure maritime safety and facilitate economic recovery from the COVID-19 pandemic. Circular Letter No.4204/Add.30. 11 September. Available at https://wwwcdn.imo.org/localresources/en/MediaCentre/HotTopics/Documents/COVID%20CL%204204%20adds/Circular%20Letter%20No.4204Add.30%20Joint%20Statement%20Seafarers.pdf.

IMO (2020b). Coronavirus (COVID-19) – Outcome of the International Maritime Virtual Summit on Crew Changes organized by the United Kingdom. Circular Letter No.4204/Add.24. 13 July. Available at https://wwwcdn.imo.org/localresources/en/MediaCentre/HotTopics/Documents/COVID%20CL%204204%20adds/Circular%20Letter%20No.4204-Add.24%20-%20Coronavirus%20(Covid-19)%20

-%20Outcome%20Of%20The%20International%20MaritimeVirtual%20Summit%20On%20Crew-%20Change.pdf.

IMO (2020c). Recommended action to facilitate ship crew change, access to medical care and seafarer travel during the COVID-19 pandemic. Resolution MSC.473. 21 September. Available at https://www.intercargo.org/wp-content/uploads/2020/09/Resolution-MSC.473ES.2-on-Crew-Change.pdf.

IMO (2020d). Coronavirus (COVID-19) – "No crew change" clauses in charterparties. Circular Letter No.4204/Add.36/Rev.1. 23 December. Available at https://wwwcdn.imo.org/localresources/en/MediaCentre/HotTopics/Documents/COVID%20CL%204204%20adds/CL.4204-Add.36%20no%20crew%20change%20clause.pdf.

IMO (2021a). Industry Recommended Framework of Protocols for ensuring safe ship crew changes and travel during the Coronavirus (COVID-19) pandemic. MSC.1/Circ.1636/Rev.1. 22 April. Available at https://wwwcdn.imo.org/localresources/en/MediaCentre/HotTopics/Documents/MSC%201636%20protocols/MSC.1-Circ.1636%20-%20Industry%20Recommended%20Framework%20Of%20Protocols%20For%20Ensuring%20Safe%20Ship%20Crew%20Changes%20And%20Travel.pdf.

IMO (2021b). COVID-19 crew change crisis still a challenge – IMO Secretary-General. 19 March. Available at https://www.imo.org/en/MediaCentre/PressBriefings/pages/Crew-change-COVID-19.aspx.

IMO (2021c). Coronavirus (COVID-19) – Joint Statement calling on all Governments to prioritize COVID-19 vaccination for seafarers and aircrew. Circular Letter No.4204/Add.38. 25 March. Available at https://wwwcdn.imo.org/localresources/en/MediaCentre/Documents/Circular%20Letter%20No.4204-Add.38%20-%20Coronavirus%20(Covid-19)%20-%20Joint%20Statement%20Calling%20On%20All%20GovernmentsTo%20Prioritize%20Covid-19...%20(Secretariat).pdf.

IMO (2021d). Seafarers and aircrew need priority COVID-19 vaccination. 26 March. Available at https://www.imo.org/en/MediaCentre/PressBriefings/pages/vaccination-UN-joint-statement.aspx.

IMO (2021e). Report of the Maritime Safety Committee on its 103rd session. MSC 103/21/Add.1, Annex 15. 14 June.

IMO (2021f). Coronavirus (COVID-19) – Designation of seafarers as key workers. Circular Letter No.4204/Add.35/Rev.7 20 May. Available at https://wwwcdn.imo.org/localresources/en/MediaCentre/HotTopics/Documents/COVID%20CL%204204%20adds/Circular%20Letter%20No.4204-Add.35%20-%20Coronavirus%20%28Covid-19%29%20-%20Designation%20Of%20Seafarers%20As%20Key%20Workers.pdf.

INTERTANKO (2020). Outbreak Management Plan: COVID-19 (version 3). September. Available at https://www.intertanko.com/info-centre/coronavirus-resources.

ITF (2020). ITF and JNG Joint Statement: On Seafarers' Rights and the Present Crew Change Crisis. 5 October. Available at https://www.itfglobal.org/en/news/itf-and-jng-joint-statement-seafarers-rights-and-present-crew-change-crisis.

Maritime Industry Authority (2020). A Letter to All Filipino Seafarers Around the World. 13 April. Available at https://marina.gov.ph/2020/04/13/a-letter-to-all-filipino-seafarers-around-the-world.

OHCHR (2015). OHCHR, 2015, Human Rights in the 2030 Agenda for Sustainable Development. Available at http://www.ohchr.org/Documents/Issues/MDGs/Post2015/HRAndPost2015.pdf.

Philippine Statistics Authority. Available at https://psa.gov.ph/national-accounts/base-2018/data-series.

Premti A, Asariotis R (2021). Facilitating crew changes and repatriation of seafarers during the COVID-19 pandemic and beyond. 2 March. Available at https://unctad.org/news/facilitating-crew-changes-and-repatriation-seafarers-during-covid-19-pandemic-and-beyond.

Safety4Sea (2021). Belgium vaccinates foreign seafarers onboard ships. 27 July. Available at https://safety4sea.com/belgium-vaccinates-foreign-seafarers-onboard-ships/.

The World Bank. Available at https://data.worldbank.org. Philippines Overseas Employment Administration. Available at https://www.poea.gov.ph/ofwstat/ofwstat.html.

TradeWinds (2021). Lack of international vaccination strategy is failing seafarers. 30 June. Available at https://www.tradewindsnews.com/opinion/lack-of-international-vaccination-strategy-is-failing-seafarers/2-1-1032040.

UN Global Compact, et al. (2021). Maritime human rights risks and the COVID-19 crew change crisis: a tool to support human rights due diligence. Available at https://ungc-communications-assets.s3.amazonaws.com/docs/publications/Maritime-Human-Rights-Risks-and-the-COVID-19-Crew-Change-Crisis.pdf.

UNCTAD (2020a). *Review of Maritime Transport 2020*. UNCTAD/RMT/2020. Available at https://unctad.org/system/files/official-document/rmt2020_en.pdf.

UNCTAD (2020b). Coronavirus: Let's keep ships moving, ports open and cross-border trade flowing. 25 March. Available at https://unctad.org/news/coronavirus-lets-keep-ships-moving-ports-open-and-cross-border-trade-flowing.

UNCTAD (2020c). COVID-19: A 10-point action plan to strengthen international trade and transport facilitation in times of pandemic – UNCTAD Policy Brief No. 79. Available at https://unctad.org/webflyer/covid-19-10-point-action-plan-strengthen-international-trade-and-transport-facilitation.

UNCTAD (2020d). UNCTAD-IMO Joint Statement in support of keeping ships moving, ports open and cross-border trade flowing during the COVID-19 pandemic. 8 June. Available at https://unctad.org/system/files/official-document/osg_2020-06-08_stat01_en.pdf.

UNCTAD (2020e). Technical Note on Port Responsiveness in the fight against the "invisible" threat: COVID-19. Available at https://tft.unctad.org/ports-covid-19/.

UNCTAD (2020f). UNCTAD – Repositories of measures on cross-border movement of goods and persons. Available at https://etradeforall.org/news/unctad-repositories-of-measures-on-cross-border-movement-of-goods-and-persons/.

WHO (2020a). Promoting public health measures in response to COVID-19 on cargo ships and fishing vessels. 25 August. Available at https://www.who.int/publications/i/item/WHO-2019-nCoV-Non-passenger_ships-2020.1.

WHO (2020b). WHO SAGE Roadmap For Prioritizing Uses Of COVID-19 Vaccines In The Context Of Limited Supply. 13 November. Available at https://www.who.int/publications/i/item/who-sage-roadmap-for-prioritizing-uses-of-covid-19-vaccines-in-the-context-of-limited-supply.

WHO (2020c). Public health surveillance for COVID-19: interim guidance. 16 December. Available at https://www.who.int/publications/i/item/who-2019-nCoV-surveillanceguidance-2020.8.

WHO (2021a). Roadmap to improve and ensure good indoor ventilation in the context of COVID-19. 1 March. Available at https://www.who.int/publications/i/item/9789240021280.

WHO (2021b). Policy considerations for implementing a risk-based approach to international travel in the context of COVID-19. 2 July. Available at https://apps.who.int/iris/handle/10665/342235.

WHO (2021c). Technical considerations for implementing a risk-based approach to international travel in the context of COVID-19: interim guidance: annex to: Policy considerations for implementing a risk-based approach to international travel in the context of COVID-19. 2 July. Available at https://apps.who.int/iris/handle/10665/342212.

Legal and regulatory developments and the facilitation of maritime trade

This chapter summarizes important recent international legal and regulatory developments. It also covers maritime trade and transport facilitation issues, particularly those related to COVID-19 which has created many problems for clearing goods through ports, but also created opportunities for new and smart solutions.

Many of the latest innovations in maritime transport involve online and automated systems that raise concerns about cybersecurity. However, shipowners and operators can also take advantage of recently adopted guidelines on how to maintain cybersecurity in their companies and onboard ships, taking into account the requirements of IMO, and other relevant guidelines.

The COVID-19 pandemic has highlighted many systemic weaknesses, including delays in documentation and related problems, which could provide an impetus for the more widespread use of secure electronic solutions that are already available and accepted by the market. Related work at UN bodies, including the United Nations Commission on International Trade Law (UNCITRAL), is also underway, to explore the possibility of developing a negotiable transport document or electronic record.

In addition, the industry is conducting trials on maritime autonomous surface ships (MASS). In May 2021, the IMO Maritime Safety Committee (MSC) completed a regulatory scoping exercise. A number of high-priority issues, cutting across several legal instruments, remain to be addressed at a policy level to determine future work.

In June 2021, the IMO adopted amendments to Annex VI of the MARPOL Convention aimed at reducing carbon intensity of ships and including targets for energy efficiency, to further reduce GHG emissions from shipping. The industry is also planning an International Maritime Research and Development Board, a non-governmental body funded by a $2-per-ton-levy on shipping fuel. Other important regulatory developments relate to the ship-source pollution control and environmental protection measures, including shipping and climate change mitigation and adaptation; air pollution, in particular sulphur emissions; oil pollution from ships; ballast water management; and biofouling.

Finally, the chapter addresses maritime trade and transport facilitation. This includes the Trade Facilitation Agreement of the World Trade Organization and recent amendments to the FAL Convention related to digitalization, concluding with a section on UNCTAD's ASYHUB Maritime system.

Legal and regulatory developments and the facilitation of maritime trade

 Development of maritime autonomous surface ships (MASS) technology and trials, as well as related regulatory responses, are advancing

 With increasing automation and digitalization, there is a growing need to effectively protect shipping assets and technology from cyber threats

 Climate-change adaptation and resilience-building for seaports is becoming an increasingly urgent challenge, especially for vulnerable developing countries that are at high and growing risk of climate change impacts

 IMO Member States agree on new mandatory regulations to further reduce GHG emissions from international shipping

 Digitalization and automation of trade procedures such as Maritime Single Windows are catalysts for more efficient and paperless compliance processes at ports

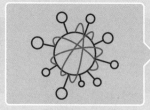 Multilateral Agreements such as the WTO TFA and the IMO FAL Convention provide solid international standards to build automated systems while ensuring interconnectivity and interoperability

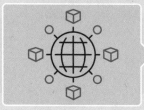 Building resilient and efficient logistic supply chains requires public-private dialogue. Cooperation from businesses involved in maritime trade and port operations through National Trade Facilitation Committees foster successful trade reforms

A. TECHNOLOGICAL DEVELOPMENTS IN THE MARITIME INDUSTRY

1. Ensuring maritime cybersecurity

The maritime sector is increasingly structured around online and automated systems. These are appearing in shipping, port operations, offshore infrastructure, and digital commercial transactions. Online platforms and information systems have many advantages but also expose the industry to new and unforeseen threats and vulnerabilities, notably the risk of cyberattacks (British Ports Association, 2020). In response, in recent years the IMO has adopted number of international instruments and developed tools for assessing the cybersecurity risks and vulnerabilities of the international maritime sector and strengthening the resilience of vital systems of shipping companies, ships and ports.[1]

More recently, the industry organization BIMCO issued 'Guidelines on Cyber Security on board Ships – fourth version' (BIMCO et al., 2021).[2] Taking account of IMO guidelines and the US National Institute of Standards and Technology (NIST) framework, the guidance specifies, for example, that company plans and procedures for cyber-risk management should be incorporated into existing security and safety risk management requirements contained in the International Safety Management Code (ISM) Code and International Ship and Port Facility Security (ISPS) Code.

According to the BIMCO guidelines, enterprises should:

- *Identify cybersecurity threats* – to the ship, both external and internal, including those posed by inappropriate use, and poor cybersecurity practices.

- *Identify vulnerabilities of assets within the company* – and develop inventories of onboard systems with direct and indirect communications links. Everyone concerned should understand the consequences of cybersecurity threats and the capabilities and limitations of existing protection measures.

- *Assess risk exposure, and vulnerabilities* – and the potential for such vulnerabilities being exploited.

- *Develop protection and detection measures* – to reduce the likelihood of vulnerabilities being exploited and the potential impact.

- *Establish response plans* – including contingency plans to respond to cyber-risks, and tackle the effects of potential attacks on ship safety and security.

- *Respond and recover* – from any cyber security incidents using the contingency plan, then report on the effectiveness of the response plan, update it, and reassess threats and vulnerabilities (BIMCO et al., 2021).

The maritime industry is increasingly taking action against these threats, but much remains to be done. Maintaining effective cybersecurity is not easy. It requires collaborative, top-down approaches that engage senior management, combined with bottom-up approaches working with other staff to identify vulnerabilities and risks unique to each operational environment – all the while balancing and managing such risks within acceptable limits.

Implementing cybersecurity helps to protect shipping assets and technology from cyber-threats and makes economic sense. But inaction could also result in consequences. Shipowners who fail to comply with the IMO requirements risk having their ships detained by port control authorities – though enforcement should be uniform and equitable.

Failure to address cybersecurity may also result in potential contractual liability. Cyberattacks can cause damage, loss or misappropriation of cargos, with implications for liability in the context of contracts for the carriage of goods by sea. Arguably, a shipowner's obligation to exercise due diligence, and provide a seaworthy vessel before and at the beginning of the voyage (see Art. III, r. 1 and IV, r. 1, Hague-Visby Rules[3]), may also include an obligation to conduct regular cybersecurity risk assessments, and address

---

[1] For further information, and an overview of IMO, ISO, EU, US and industry cybersecurity guidance, see UNCTAD, 2020a, chapter 5. See also IMO, 2021a.

[2] Other available guidelines include the Digital Container Shipping Association's Implementation Guide for Cyber Security on Vessels v1.0 (DCSA, 2020), based on version 3 of the industry guidelines (BIMCO et al., 2018), and the US NIST framework (NIST, 2018). While their target audience is the container industry, other segments of shipping may also find them useful. In addition, the International Association for Classification Societies (IACS) has issued a recommendation (IACS 2020), which applies to newbuild ships only, but can also serve as guidance for existing ships.

[3] https://www.jus.uio.no/lm/sea.carriage.hague.visby.rules.1968/portrait.pdf.

risks and reduce vulnerabilities through safety management systems, in accordance with IMO and industry guidance.

For ports, BIMCO and other maritime NGOs have invited public and private stakeholders to help create global digital ISO standards to facilitate the digital exchange of data, particularly in light of the new urgency brought about by the COVID-19 pandemic and increasing demand (BIMCO, 2021).

2. Maritime autonomous surface ships

The use of maritime autonomous surface ships (MASS) could increase safety and improve environmental performance, and accelerate decarbonization. Various countries are moving ahead fast with this technology and currently have MASS commercial projects at the stage of advanced testing and trialling (Gard, 2020; Yara, 2020).

To enable the safe, secure, and environmentally sound operation of MASS within the existing IMO instruments, the IMO has been considering amending its regulatory framework (IMO, 2017, para. 20.2). These issues are also being considered by the academic community, industry, and governments. In 2017 the IMO Maritime Safety Committee (MSC), embarked on a regulatory scoping exercise which it completed in May 2021. This should also help progress related discussions in other IMO Committees namely LEG, MEPC and FAL (see also UNCTAD, 2019).

For each provision under its purview the MSC considered whether MASS could be regulated by either: equivalences as provided by the instruments or developing interpretations; and/or amending existing instruments; and/or developing a new instrument; or none of the above.[4]

The committee highlighted high-priority issues that cut across several instruments. An immediate concern is terminology – including the definition of a MASS and clarifying terms such as "master", "crew" and "responsible person" which should be agreed internationally in cooperation with the ISO. The MSC has also considered the function and operations of the remote-control station or centre, and the possible designation of a remote operator as a 'seafarer'. The committee has identified other issues across several safety treaties related to: manual operations and alarms on the bridge; actions by personnel, such as firefighting, cargo stowage and securing and maintenance; watchkeeping; search and rescue; and the information required to be on board for safe operation.

The MSC noted that the best way to address these gaps and themes would be to proceed in a holistic manner. This should result in a MASS instrument/Code whose goals, functional requirements, and corresponding regulations, are suitable for all four degrees of autonomy. For further work it will be important to establish a joint MSC/LEG/FAL working group, but in the meantime these committees can liaise on common issues and align any future work (IMO, 2021b).

In July 2021, the IMO Legal Committee completed its scoping exercise, concluding that MASS could be accommodated within the existing LEG conventions without the need for major adjustments or a new instrument. Some conventions can accommodate MASS as drafted, though others may require additional interpretations or amendments (IMO, 2021c).

B. REGULATORY DEVELOPMENTS RELATING TO INTERNATIONAL SHIPPING, CLIMATE CHANGE AND OTHER ENVIRONMENTAL ISSUES

1. IMO action on greenhouse gas emissions

In April 2018 the IMO adopted its initial strategy on reducing greenhouse gas (GHG) emissions from ships (see IMO, 2018, annex 1; UNCTAD, 2019). This envisages emissions peaking as soon as possible and by 2050 falling to at least 50 per cent below the 2008 level, with the aim of being phased out entirely. By 2030 the target is to reduce the carbon intensity of international shipping by at least 40 per cent of the 2008 level (IMO, 2020a).

In June 2021, in line with the IMO initial strategy, the MEPC adopted new mandatory regulations as amendments to Annex VI of the MARPOL Convention. These build on earlier efficiency requirements and aim to cut the carbon intensity of existing ships, and further reduce GHG emissions from shipping – requiring operators to measure the energy efficiency of all ships and meet specified targets.

---

[4] The outcome of the MSC's regulatory scoping exercise, as approved by the Committee, including the full analysis of treaties, can be found as an annex to the report on its 103rd session (IMO, 2021b).

For this purpose, operators can use a new Energy Efficiency Existing Ship Index (EEXI), along with a new operational carbon intensity indicator (CII) – a dual-track approach that will enable them to address both technical and operational measures. The EEXI measures the energy efficiency of the ship compared to a baseline and should be calculated for ships of 400 GT and above, in accordance with values set for ship types and size categories. Ships are required to reduce the EEXI by a specified percentage of the baseline.

Ships of 5,000 GT are already required to collect data on fuel oil consumption. Now they must also bring their operational carbon intensity within a specific level, document and verify their CII against the required value, and record this in the Ship Energy Efficiency Management Plan (SEEMP). This should result in a performance rating of A, B, C, D or E – corresponding to major superior, minor superior, moderate, minor inferior, or inferior. A ship rated D for three consecutive years, or E, would have to submit a plan for corrective action, to show how the required rating (C or above) would be achieved. Administrations, port authorities and other stakeholders are encouraged to provide incentives to ships rated A or B.

These amendments are expected to enter into force on 1 November 2022, with the requirements for EEXI and CII certification coming into effect from 1 January 2023. This will allow the first annual reporting on carbon intensity to be completed in 2023, with the first rating given in 2024. For its part, the IMO is to review the effectiveness of the implementation by 1 January 2026 and, if necessary, adopt further amendments. To support the implementation, the MEPC has also adopted related guidelines.

The GHG reduction candidate measures considered at IMO need to undergo an initial assessment of their impact on States, based on the procedure adopted in 2019 (MEPC.1/Circ.885). The procedure also states that proposed measures, including the latest measures adopted, need to undergo a comprehensive impact assessment before adoption if required by the Committee. To support this process, UNCTAD has been collaborating with the IMO on an expert review of the impact assessments submitted to ISWG-GHG 7, as well as the final comprehensive impact assessment of the short-term combined measures submitted to the 76th session of MEPC (UNCTAD, 2021a; see also chapters 2 and 4 for a discussion of the outcomes).

The 75th and 76th sessions of the MEPC also discussed an industry-led proposal for a non-governmental International Maritime Research and Development Board (IMRB), funded by a mandatory $2 per-tonne levy on ship fuel. The MEPC also considered mid- and long-term measures, including market-based measures, and a work plan for further cutting GHG emissions from shipping, in line with the initial IMO strategy (IMO, 2021d). Further consideration of the proposals should take place during ISWG-GHG 10 in October 2021.

2. Adapting transport infrastructure to the impacts of climate change

In August 2021, less than three months before COP26 in Glasgow in November 2021, the Intergovernmental Panel on Climate Change issued its 6th Assessment Report (AR6) (IPCC, 2021). This was the first comprehensive review of the science of climate change since 2013 and gave clear warnings of increasingly extreme heatwaves, droughts, and flooding that could have devastating consequences, making effective adaptation action a matter of increasing urgency. AR6 projects that, depending on scenario, the mean global temperature increase of 1.5°C relative to pre-industrial times is likely to be reached by 2040; and if emissions are not slashed in the next few years this threshold may be reached even earlier. Nevertheless, these impacts can be avoided if the world acts quickly with essential measures for adaptation and mitigation (IPCC 2018; IPCC 2019; IPCC 2021).

Adaptation will be particularly important for seaports. Ports are exposed to various climate hazards, including heat waves, extreme winds and precipitation, as well as a rise in mean sea level and associated extreme sea-levels (IPCC, 2019). This consideration, which is of particular importance from the perspective of developing countries, was highlighted again in October 2020, at the eighth session of the UNCTAD Multi-year Expert Meeting on Transport, Trade Logistics and Trade Facilitation which focused on "Climate change adaptation for seaports in support of the 2030 Agenda for Sustainable Development" (UNCTAD, 2020c) (UNCTAD, 2020d). Effective adaptation will need to be underpinned by strong legal and regulatory frameworks, along with strategies, policies and plans to reduce vulnerability. For this purpose, stakeholders will need the appropriate standards, guidance and tools.

One of the outcomes of COP22 was the Marrakech Partnership for Global Climate Action[5], which is designed to provide a strong foundation for how the UNFCCC process will catalyse and support climate action. This has produced the 'Climate Action Pathway for Transport' which includes recommendations for 'Resilient transport systems, infrastructure and vehicles', with milestones towards 2050 (for 2025, 2030

---

[5] See https://unfccc.int/climate-action/marrakech-partnership/reporting-and-tracking/climate_action_pathways.

and 2040) (UNFCCC, 2021a and 2021b). By 2025, all new transport infrastructure, systems and, where necessary vehicles, should be climate-resilient to at least 2050; by 2030, that should extend to all critical transport infrastructure and systems. By 2040, all critical infrastructure and systems should be climate-resilient to at least 2100 (UNFCCC, 2021b).

Translating this timely ambition into action will require a major acceleration of efforts. For its part, in 2021 the EU issued its Climate Change Adaptation Strategy, which aims for a climate-resilient EU by 2050 – "by making adaptation smarter, more systemic, swifter, and by stepping up international action" (European Commission, 2021). The EU has also adopted a new Climate Law, which entered into force on 29 July 2021 (European Union, 2021). This aims for EU climate neutrality by 2050 and by 2030 to reduce domestic net greenhouse gas emissions by at least 55 per cent of their 1990 levels. In addition, the new law envisages "continuous progress in enhancing adaptive capacity, strengthening resilience and reducing vulnerability to climate change in accordance with Article 7 of the Paris Agreement" and related stocktaking, starting in 2023.

Guidance for action has also been produced by the World Association for Waterborne Transport Infrastructure (PIANC). In 2020 PIANC issued a revised version of 'Climate Change Adaptation Planning for Ports and Inland Waterways' (PIANC 2020). This covers priority actions such as: inspection and maintenance; monitoring systems and effective data management; and risk assessments, contingency plans and warning systems. It also focuses on flexible and adaptive infrastructure, systems and operations and better resilience through engineered redundancy.

Also worth noting is the new ISO standard ISO 14091:2021 – Adaptation to climate change-Guidelines on vulnerability, impacts and risk assessment (ISO, 2021). This covers vulnerability to climate change, and highlights the importance of risk assessments and of monitoring and evaluating for any organization, regardless of size, type, or nature.

In 2020 during the COVID-19 pandemic, there was a significant fall in investment in transport infrastructure.[6] However, major scaling up of investment and capacity building for developing countries will be critical to 'building back better' after the pandemic. The OECD estimates that meeting the SDGs by 2030 will require $6.9 trillion in infrastructure investment annually, (OECD, 2017). At a recent UNCTAD dialogue, SIDS representatives highlighted the urgent need for better availability/access to green and blue infrastructure financing (UNCTAD, 2021b and c). This could bring enormous economic benefits: the World Bank estimates that investing in resilient infrastructure in developing countries could bring returns of $4.2 trillion over the lifetime of new infrastructure – a $4 benefit for each dollar invested (Hallegatte S. et al., 2019).

3. Protecting the marine environment and biodiversity

Recent regulatory actions for the protection of the marine environment and conservation and the sustainable use of marine biodiversity,[7] include the following:

a) Implementing the IMO 2020 sulphur limit

Limiting SOx emissions from ships is important to improve air quality and protect both human health and the environment. On 1 January 2020 an IMO regulation entered into force that reduces the limit on the sulphur content in ship fuel oil from 3.5 to 0.5 per cent. In designated emission control areas, the limit remained even lower, at 0.1 per cent.[8] To further support enforcement, in December 2020, the MEPC adopted several amendments to MARPOL Annex VI, which will enter into force on 1 April 2022. These mainly relate to definitions and onboard sampling of the sulphur content of fuel oil, fuel verification

---

[6] According to UNCTAD, investment in transport infrastructure, power generation/distribution (except renewables) and telecommunications was down 60 per cent compared to 2019, https://unctad.org/programme/covid-19-response/impact-on-trade-and-development-2021#aTransport.

[7] As regards negotiations on a new international legal instrument under the UNCLOS on the Conservation and Sustainable Use of Marine Biological Diversity of Areas beyond National Jurisdiction, discussions on a broad range of issues, including marine genetic resources; area-based management tools, including marine protected areas; environmental impact assessments; and capacity-building and marine technology transfer, were expected to continue during the fourth session of the Intergovernmental conference on an international legally binding instrument, scheduled to be held from 23 March to 3 April 2020, but were postponed due to COVID-19 crisis (for information on discussions at earlier sessions, see UNCTAD, 2019, 2020a). The next session of the conference was scheduled to take place from 16 to 27 August 2021, but due to the COVID-19 situation, it was again postponed to the earliest possible available date in 2022, preferably during the first half of the year (see A/75/L.96).

[8] The four emission control areas are: the Baltic Sea area; the North Sea area; the North American area (covering designated coastal areas of Canada and the United States); and the United States Caribbean Sea area (around Puerto Rico and the United States Virgin Islands).

procedures, and consequent related amendments to the International Air Pollution Prevention (IAPP) certificate.

From 1 January 2020, Flag and Port State controls have had to make sure that ships comply with the 0.5 per cent sulphur limit. To do so, shipowners and charterers can adopt three different approaches:

a) Use a compliant fuel which is low enough in sulphur such as VLSFO or MGO;

b) Use alternative fuels such as liquefied natural gas (LNG), methanol, liquefied petroleum gas (LPG), hydrogen fuel cells, or biofuels which emit very small amounts of SOx; or

c) Use equivalent methods, including fitting or retro-fitting their ships with exhaust gas cleaning systems, also known as scrubbers. Scrubbers may be open loop –discharging wash water into the sea – or closed loop discharge residues to adequate reception facilities ashore.

During 2020 and the first half of 2021, implementation, primarily with the use of VLSFO, was relatively smooth, and compliant fuel oil was widely available globally (IMO, 2021e). There was some disruption by COVID-19, and several more ports and countries banned open-loop scrubber wash water discharge. Global enforcement of the new regulation was facilitated, however, by a ban on the carriage of non-compliant fuel.

Liability for compliance mainly rests with shipowners – who typically supply the fuel. In the case of charterparties, usually voyage charters, the contract may require the shipowner to warrant that the vessel complies with international rules and regulations. For time charters, on the other hand, it is the charterers who usually purchase and provide the fuel; therefore contractual provisions may shift responsibility for compliance with applicable Sulphur Content Requirements to the charterers, so the liability and the associated risk is divided between them and the shipowners, who warrant that the vessel itself is compliant. Examples of relevant clauses include the BIMCO's Marine Fuel Sulphur Content Clause for Time Charter Parties (BIMCO, 2018), and INTERTANKO's Bunker Compliance Clause (INTERTANKO, 2018). In order to increase clarity, contracting parties should consider incorporating such clauses in charterparties.

Further special regulation has been agreed for the environmental protection of Arctic waters. In June 2021, the MEPC adopted amendments to MARPOL Annex I that prohibit the use, and carriage for use of heavy fuel oil by ships in Arctic waters on and after 1 July 2024. Ships that meet certain standards on oil fuel tank protection would need to comply on and after 1 July 2029.

However, up to 1 July 2029 a Party with a coastline bordering Arctic waters may temporarily waive the requirements for ships flying its flag and operating in waters that are subject to that Party's sovereignty or jurisdiction. After that date, exemptions and waivers would no longer apply. Currently, MARPOL Annex I regulation 43 prohibits the use or carriage of heavy-grade oils on ships in the Antarctic; and under the Polar Code[9] ships are encouraged not to use or carry such oil in the Arctic. The new regulation will help protect these fragile areas further. However, its impact could be significantly reduced by the waivers and exemptions for contracting States with a coastline bordering Arctic waters, until 2029.

b) Ballast water management

One of the greatest threats to the world's oceans and a major threat to biodiversity is ships discharging untreated ballast water. This has severe consequences for public health and has environmental and economic implications for fisheries and the exploration of marine genetic resources (see also UNCTAD 2011, 2015b). In December 2020, the MEPC adopted amendments to the International Convention for the Control and Management of Ships' Ballast Water and Sediments, 2004 (the BWM Convention) which aims to prevent the introduction and proliferation of non-native species following the discharge of untreated ballast water from ships. These amendments, which are expected to enter into force on 1 June 2022, relate to the commissioning and testing of ballast water management systems and to the form of the International Ballast Water Management Certificate. As of 31 July 2021, the BWM Convention had 86 Contracting States representing 91 per cent of the GT of the world's merchant fleet.[10]

c) Biofouling

A prominent, but underestimated, source of microplastic pollution is antifouling coatings on ships (Dibke C. et al., 2021). In June 2021, the MEPC adopted amendments to the IMO Convention for the Control

---

[9] For more information, see UNCTAD, 2015a.

[10] https://wwwcdn.imo.org/localresources/en/About/Conventions/StatusOfConventions/StatusOfTreaties.pdf.

of Harmful Anti-fouling Systems on Ships, 2001 (AFS Convention)[11], to prohibit anti-fouling systems containing cybutryne. This would apply from 1 January 2023 or, for ships already using such a system, at its next scheduled renewal after 1 January 2023, but no later than 60 months following the last application to the ship of such an anti-fouling system.[12]

d) Oil-pollution from shipping

An important risk of pollution is oil spills from ships, not just from oil tankers, but also from other maritime transport – container ships, chemical carriers, general cargo ships and passenger or cruise vessels. Oil spills, and the resultant clean-up operations, can seriously affect marine and coastal environments, from both physical smothering and the effects of toxins. There are also costly and wide-ranging economic implications (Asariotis R., Premti A., 2020). The risks are particularly high for vulnerable coastal developing states and ocean economies such as SIDS that rely heavily on fisheries, aquaculture and tourism, and are being heightened by bigger vessels carrying high volumes of bunker fuel oil.

The 'Wakashio' bunker oil spill off the coast of Mauritius in 2020 demonstrated the devastating consequences of oil spills for the economies and tourism industries of coastal countries, as well as for ecosystems and biodiversity, further endangering corals, fish, and other marine life (IPCC 2018). This spill also highlighted the need for international legal instruments in this field and for all States to adopt the latest of these.

Oil spills raise serious issues of liability and of compensation, including for the costs of reinstating the environment. In this respect there is a comprehensive international regime in place on liability and compensation for oil pollution damage caused by persistent oil spills from tankers (CLC-IOPC Fund regime) (UNCTAD, 2012).[13] Unfortunately, this did not apply in the Wakashio case, as the spill was of bunker oil from a bulk-carrier, not from an oil tanker (Asariotis R., Premti A., 2020; UNCTAD 2020b).

Bunker oil spills from ships other than oil tankers are covered by the International Convention on Civil Liability for Bunker Oil Pollution Damage, 2001 (Bunkers Convention).[14] This Convention aims "to ensure that adequate, prompt, and effective compensation is available to persons who suffer damage caused by spills of oil, when carried as fuel in ships' bunkers". Modelled after the International Convention on Civil Liability for Oil Pollution Damage, 1992 (CLC), the Bunkers Convention has many similar provisions but the amount of liability may be limited (Art. 6), in accordance with any applicable national or international regime such as the Convention on Limitation of Liability for Maritime Claims (LLMC), 1976, as amended in 1996. As a result, the compensation available to claimants is significantly lower than that available under the CLC-IOPC Fund regime for oil pollution from tankers.[15] Given the continuing growth in sizes of ships of all types, the issue of liability for bunker oil spills from ships other than tankers may need to be revisited.

A Claims Manual for the Bunkers Convention

For the IOPC FUNDS, there is a Claims Manual but there is no corresponding manual for the Bunkers Convention. During its 107th session in December 2020, the IMO Legal Committee supported the development of an 'IMO Claims Manual for the Bunkers Convention' to guide national courts, claimants, shipowners and insurers in their interpretation of the Convention (IOPC FUNDS, 2019). This manual would differ from the 1992 Fund Claims Manual but should be consistent with it. The Committee agreed that, in cooperation with protection and indemnity clubs, a more detailed proposal would be taken forward on an intersessional basis, (IMO, 2020b, pg. 27). Then in July 2021 at its 108th session the Legal Committee expressed its broad support for the development of dedicated and authoritative guidance for claimants within the scope of the Convention (IMO, 2021c). Such a manual would assist claims under the Convention, but it should also reflect the needs of vulnerable coastal developing countries and SIDS, particularly on the question of limitation of liability.

---

[11] For some background information, see *Review of Maritime Transport 2020*.

[12] The Convention, which as of 31 July 2021, was in force for 91 Contracting States representing 95.93 per cent of the GT of the world's merchant fleet, defines "anti-fouling systems" as "a coating, paint, surface treatment, surface or device that is used on a ship to control or prevent attachment of unwanted organisms". It already prohibits the use of harmful organotin compounds in anti-fouling paints used on ships and establishes a mechanism to prevent the potential future use of other harmful substances in anti-fouling systems. These harmful substances include the biocide chemical compound cybutryne, for which scientific data has indicated it causes significant adverse effects to non-target organisms and the environment, especially to aquatic ecosystems, and therefore needs to be controlled.

[13] 1992 Civil Liability Convention (CLC), 1992 Fund Convention and 2003 Supplementary Fund Protocol. See further https://www.iopcfunds.org/.

[14] http://library.arcticportal.org/1616/1/6693.pdf.

[15] See, https://www.imo.org/en/About/Conventions/Pages/Convention-on-Limitation-of-Liability-for-Maritime-Claims-(LLMC).aspx.

Limitation of liability under IMO conventions

In certain circumstances a shipowner may lose the statutory right to limitation of liability under some international conventions. The IMO Legal Committee has also been discussing a unified interpretation on the relevant test for breaking the shipowner's right to limit liability (see IMO, 2019a, 2019b). In December 2020, the Committee established a remote intersessional group to draft such a unified interpretation and consider the vehicle for its adoption – which would be either the Conference of States Parties, the Assembly, or the Legal Committee. Drawing on this work, a related draft Assembly Resolution has since been finalized by the IMO Legal Committee at its 108th session and submitted for consideration by the Assembly at the end of the year (IMO, 2021c).

C. LEGAL AND REGULATORY IMPLICATIONS OF THE COVID-19 PANDEMIC

The COVID-19 pandemic is causing delays and unprecedented supply-chain disruptions that affect the performance of a wide range of contractual obligations and can lead to the need for costly litigation, involving complex jurisdictional issues in a global context. This could be on such a scale as to overwhelm some legal and administration of justice systems, with implications for global governance and the rule of law.[16]

Avoiding this outcome will require collective and coordinated action by governments and industry. This could involve, for example agreeing contractual extensions, showing restraint in pursuing legal rights and claims, and resolving disputes through mediation and arbitration, as well as strengthening formal and informal dispute resolution mechanisms and institutions. It could also involve commercial risk-allocation through standard clauses drafted to address contractual rights and obligations in the light of the circumstances associated with the pandemic.

As part of UN action in response to the COVID-19 pandemic, UNCTAD and the UN regional Commissions are currently implementing a joint technical assistance project: "Transport and trade connectivity in the age of pandemics: Contactless, seamless and collaborative UN solutions".[17] UNCTAD is leading several of these components, including work on the international commercial transport and trade law implications of the pandemic, and has already published two briefing notes: one on Cargo Claims, (UNCTAD, 2021d; the other on International Sale of Goods (UNCTAD, 2021e)). These highlight some of the complex commercial law issues and implications to encourage discussions between the affected parties and consider appropriate measures for future agreements.

One issue which has clearly come to the market's attention is that of delays in documentation. This may provide an impetus for more commercial parties to adopt secure electronic solutions that are already available and have been accepted by the market. However, with increasing reliance on electronic interactions, they will also have to manage any associated cyber-risks and enhance their cybersecurity systems.

Lessons learnt from the global pandemic should generally encourage carriers, insurers, and cargo interests to take leaps forward and make the best use of technology, both to minimize disruption, and to allocate fairly any commercial risks that arise from unforeseen events beyond the control of the contracting parties. Trade associations can help in this respect by devising standard form terms for inclusion into commercial contracts. In addition, governments and policymakers should consider temporary financial support to avoid widespread business failure and protect the essential flow of goods across all trade routes.

D. OTHER LEGAL AND REGULATORY DEVELOPMENTS AFFECTING TRANSPORTATION

1. Combating fraudulent registration and registries

In 2019, following reports by several members on the fraudulent use of their flags, the IMO Legal Committee, agreed on measures to prevent fraudulent ship registration and registries (UNCTAD, 2019). The Committee supported the development of a comprehensive database of registries to be held on the publicly available contact points module of the IMO Global Integrated Shipping Information System.

---

[16] Note in this context also SDG 17, which focuses on partnership for the goals, and SDG 16 on peace, justice, and strong institutions.

[17] https://unctad.org/project/transport-and-trade-connectivity-age-pandemics.

This would contain the names and contact details of national governmental bodies or authorized/delegated entities in charge of the registration of ships, as well as other relevant information. The Committee also approved best practices to combat fraudulent registration and registries of ships, and established an intersessional correspondence group to consider various proposals in greater detail (IMO, 2019b). This group, in which UNCTAD participated, has since prepared a draft Resolution on "Encouragement of Member States and all relevant stakeholders to promote actions for the prevention and suppression of fraudulent registration and fraudulent registries, and other fraudulent acts in the maritime sector". This was finalized by the IMO Legal Committee at its 108th session in July 2021 and submitted to the IMO Assembly, for consideration in December 2021 (IMO, 2021c). The intersessional group had also proposed future work on a corresponding IMO study, which was agreed upon by the IMO Legal Committee. It should be noted that there is already an International Convention on the Registration of Ships, 1986,[18] which provides some safeguards against fraudulent ship registration, and was adopted under the auspices of UNCTAD, but it has not entered into force.

2. Multimodal transport discussions at UNCITRAL and ESCAP

Multimodal transport can be a key driver of sustainable development, by enabling existing capacities and infrastructure to be used more effectively and promoting a better balance between transport modes across supply-chains. However, the international legal framework is lagging behind. Despite numerous attempts, no uniform legal regime on multimodal transport has entered into force internationally (UNCTAD, 2003). Instead, the existing framework consists of a complex jigsaw of international conventions designed for unimodal carriage, regional and sub-regional agreements, national laws, and standard term contracts. This is associated with a lack of legal certainty and a need for costly evidentiary enquiries and litigation.

ESCAP – Harmonizing multimodal legal frameworks in Asia and the Pacific

In August 2020, a ESCAP Expert Group Meeting, in which UNCTAD participated, discussed options for harmonizing the legal framework for multimodal transport at the regional level. The Expert Group requested a more detailed analysis of the advantages, disadvantages and specificities of each option – including the level of commitment needed, the timelines for completion and the potential for causing additional fragmentation or legal conflicts. (ESCAP, 2020).

In March 2021 this analysis was discussed by a second Expert Group Meeting (ESCAP, 2021). Several participants highlighted the value of a single comprehensive legal instrument, but the meeting concluded that would be more practical to take a step-by-step approach. This included consideration of the following possibilities:

   i. Tailor-made legal solutions addressing specific modal interfaces.
   ii. A single transport document that could serve as evidence of a contract.
   iii. Digitalization of consignment notes.
   iv. A framework agreement together with soft law solutions.
   v. Solutions building on existing infrastructure networks and agreements, such as an instrument on multimodal transport operations envisaged under the Intergovernmental Agreement on Dry Ports.

The secretariat was requested to take these elements into account and to provide relevant background material for the next meeting.

UNCITRAL – Negotiable multimodal transport documents

In July 2019, at the 52nd session of UNCITRAL, the Government of China presented a proposal on possible future work by UNCITRAL to develop a legal framework for railway consignment notes. This noted that railway transportation had some advantages, such as shorter distances, greater speed, and less vulnerability to weather. However, unlike ocean bills of lading which were used for maritime transport, international railway consignment notes did not serve as documents of title and were not used for the settlement and financing of letters of credit. UNCITRAL considered that the proposal could be of practical significance for world trade, and particularly for the economic growth of developing countries. However, given the complexity of the issues, the Commission decided, as a first step, to request the Secretariat to coordinate with other relevant organizations and conduct research on the legal issues related to the use of railway or other consignment notes, (UNCITRAL, 2019, paras. 216 –217).

---

[18] Text is available at UNCTAD's website, https://unctad.org/topic/transport-and-trade-logistics/policy-and-legislation.

Expert Group meetings were held in 2019 and 2020, and in May 2020 their conclusions were presented to the 53rd annual session of UNCITRAL (UNCITRAL, 2020a). The Commission recognised the value of electronic transport documents, particularly for the new supply chain and logistics models expected to develop following the COVID-19 disruption and requested the secretariat to start preparatory work, in close coordination and cooperation with relevant international organizations, on a new international instrument on multimodal negotiable transport documents that could be used for contracts not involving carriage by sea (UNCITRAL, 2020b, para.16(e)).

In February 2021, there was a Third Expert Group Meeting on a 'New International Instrument on Negotiable Multimodal Transport Documents', with the participation of international organizations, including UNCTAD, as well as practitioners and academia. In April 2021, an open webinar on 'International experiences with the dematerialization of negotiable transport documents' was held (UNCITRAL, 2021a). At its 54th session in July 2021, UNCITRAL welcomed the preparatory work and confirmed its strong interest in the project. The Commission agreed that "the primary purpose of a new international instrument should be to ensure legal recognition of a medium neutral negotiable transport document in different modes of transport and that, for that purpose, it was desirable to focus first on negotiable transport documents and subsequently consider whether other types of transport documents accepted by banks for documentary credit should also be encompassed". The Commission also agreed on the need for proper coordination and interface with the liability regimes provided under existing conventions on international carriage of goods by various modes and invited the secretariat to continue its preparatory work in close coordination with other organizations currently working on or exploring solutions to enable the use of a negotiable transport document in the rail plus or other multimodal context, as well as other organizations with relevant expertise, or representing relevant industries (UNCITRAL, 2021b).

Given the broad substantive scope of the proposed future legal instrument, public and private stakeholders both in multimodal transport and in all the different modes are encouraged to participate in any related further work. For small traders in developing countries, a key concern will be adequate liability for cargo loss or damage. UNCTAD will continue to participate in any related work under the auspices of UNCITRAL.

3. Status of conventions

A number of international conventions in the field of maritime transport have been prepared or adopted under the auspices of UNCTAD. During the current reporting period, only the status of the Hamburg Rules changed, with one additional accession (see https://unctad.org/webflyer/review-maritime-transport-2021). For additional information, see https://unctad.org/ttl/legal. For official status information, see the United Nations Treaty Collection, available at https://treaties.un.org.

E. MARITIME TRANSPORT WITHIN THE WTO TRADE FACILITATION AGREEMENT

Implementing trade and transport facilitation procedures efficiently, and in line with international guidelines reduces time and costs, and makes for more agile logistics supply chains. This will involve simplifying maritime and trade procedures, and integrating new technologies in trade and transport facilitation so as to standardize and harmonize for cross border trade in goods.

The COVID-19 crisis has highlighted the many national regulations and administrative bottlenecks involved in the emergency supply of medical equipment, drugs – as exemplified by the ongoing vaccine supply chain. Minimizing disruption in the logistics supply chains, including maritime transport, will mean extending international frameworks, building more public-private partnerships, and further digitalizing trade facilitation.

Such reforms will rely on harmonized international frameworks such as the WTO TFA and the IMO FAL Convention. These instruments, which provide governments with guidance and incentives in reforming trade facilitation measures, are paving the way for digitalization, transparency, and rationalization of administrative formalities. They already serve as the bases for many bilateral and regional trade facilitation agreements, and other initiatives are emerging as complementary building blocks.

1. Implementation of the WTO TFA

The WTO Trade Facilitation Agreement aims to boost the speed and efficiency of cross-border trade procedures through 36 measures and covers four areas: transparency, fees and formalities, customs cooperation, and transit. As of July 2021, 154 WTO members have ratified the TFA, meaning that 94 per cent of WTO Members apply the agreement on a most-favoured-nation basis.

However, the agreement is not being implemented by all members – only by 71 per cent of developing countries and by 36 per cent of LDCs. The reality on the ground may even be less positive, as it is not sure that countries fully comply in practice with their notified implementation schedules.

Trade facilitation makes ports and shipping more efficient. Those developing countries and LDCs that implement the TFA tend to have a higher turnover of container ships at port. This is evident from the UNCTAD Liner Shipping Connectivity Index, which shows that 13 per cent of the variance of the time that container ships spend in port can be statistically explained by differences in TFA implementation (UNCTAD, 2016).

To help developing countries and LDCs implement the agreement, the TFA provides for Special and Differential Treatment (SDT) through which those countries can acquire the necessary capacity. To benefit from SDT, developing countries and LDCs need to define their needs for technical assistance and capacity building (TACB). As of July 2021, 119 developing and least developed members had notified their intention to use the SDT provisions.

Recipients of TACB have made progress in implementing TFA commitments. For LDCs that have received TACB support, OECD indicators and WCO-Time Released Study data reveal substantial reductions in customs clearance times. The progress is especially evident in transparency on customs rules and regulations, customs automation, and in the timely release and clearance of goods (OECD/WTO, 2019).

2. Measures related to maritime transport

The TFA presents the regulatory requirements for the release and clearance process of export, import and transit operations and covers procedures linked to customs clearance and to standards and controls from other border agencies (Bureau of Standards, Ministry of Agriculture, etc.). Article 7 of the TFA addresses the Release and Clearance of goods, including customs operations such as pre-arrival processing, risk management, and trade facilitation measures for authorized operators. Article 10 on Formalities connected with Importation, Exportation and Transit, addresses the relations between border agencies and the business community, and includes provisions for single window implementation, and the use of international standards and of customs brokers. Finally, articles 8 and 12 cover Border Agency and Customs Cooperation.

Some provisions are more fully implemented than others (WTO, 2021). The higher implementation rates are those for the use of customs brokers at 87 per cent, for pre-arrival processing at 74 per cent and electronic payments at 69 per cent but other provisions involving IT infrastructure such as the single window are lower at 45 per cent. Only 59 per cent implement Article 8 on Border Agency Cooperation.

The value of the TFA is demonstrated by the World Bank Logistics Performance Index (LPI) which shows that implementation of trade facilitation measures is positively correlated with logistics performance, with the greatest benefits from Article 1 on Publication, Article 6 on Fees and Charges, Article 8 on Border Agency Cooperation and Article 10 on Formalities (UNCTAD, 2016).

3. The value of public-private dialogue

Any successful trade reform relies on cooperation between public administrations and the business community. With trust and dialogue among stakeholders, the trade ecosystem can develop sustainably, and public reforms can respond to the needs of the trader community. This principle is embedded in a number of measures in the TFA – on border agency cooperation, customs cooperation, consultations, and the opportunity for the private sector to comment before adopting a legal text.

The most important component in this context is article 23.2 on the obligation to set up in each country a National Trade Facilitation Committee (NTFC). The NTFC should comprise public and private stakeholders who can devise a coherent and coordinated strategy and champion and drive the trade facilitation agenda. NTFCs may gather all border agencies, business associations, freight forwarders associations, as well as the port authorities, agencies and private sector stakeholders working on maritime trade. According to an UNCTAD survey, 40 per cent of the NTFC members come from the private sector (Ugaz, 2019).

In Kenya, for example, the NTFC has set up a Technical Working Group on the Mombasa Port Charter which includes the Kenya Port Authority. In Namibia, the NTFC comprises the Namibia Port Authority as well as the Walvis Bay Port users' association. This public-private dialogue proved useful for defining policies, improving consultations, and resolving conflict, and during the COVID-19 crisis has been used to coordinate emergency guidelines for supplies coming through ports.

Public-private dialogue and inter-agency cooperation are often manifested in the port community system (PCS) as prescribed in TFA Article 8 and Single Window, Article 10.4. The PCS is the electronic exchange platform that interfaces with existing IT systems within a port environment, including all the stakeholders, private and public. In the Port of Valencia, Spain the PSC provides for the electronic exchange of supply chain information for B2B, B2G and G2B. Recently, these systems have started to link up internationally with port-to-port data exchange– facilitated by the International Port Community Systems Association Network of Trusted Networks. In addition to pre-arrival and arrival processing this enables greater transparency in the supply chain through track and trace.

Another critical issue for public-private dialogue is the safety and well-being of workers. Ports and other actors can for example, cooperate to improve crew changeover processes and ensure standards of procedure and risk-management protocols at the national level so that imperatives of operational continuity do not compromise the safety and well-being of workers. This issue has also come to the fore during the pandemic when seafarers have suffered from blockades on ships for several months and from loss of employment and were often in desperate conditions.

The benefit of public-private cooperation has been demonstrated in the 'landlord port' system. In this case, border agencies deal with regulatory policies and administer the supply chain while the private sector oversees the handling and storage of shipments as well as the maintenance of port terminals. This allows the government to upgrade its systems for customs clearance and other regulatory treatments of goods while the business sector can improve hard infrastructure, thus boosting the port competitiveness.

4. Improving technology and extending digitalization

Trade facilitation is steadily being transformed by new technology. The TFA encourages smart solutions in the clearance of goods – as with Article 1.2 on information available through the internet, Article 10.4 on the electronic single window, or Article 7.2 on electronic payments.

The electronic single window (eSW) has revolutionized supply chains by interconnecting border agencies, traders, and logistics providers on the same IT platform. It provides a single point of submission for trade documents and information and allows border agencies to share documents and data electronically and establish common procedures for processing and control.

Rwanda, for example, has built the Rwanda Electronic Single Window (ReSW) using UNCTAD's Automated System for Customs Data (ASYCUDA). Since its introduction in 2012, the ReSW has connected approximately 20 government agencies and now provides more than 12 single window services and applications. Since 2020, new Partner Governmental Agencies like the Rwanda Agriculture and Livestock Inspection and Certification Services and the National Agricultural Export Development Board have been benefiting from automated applications in the single window system. In 2014 alone, the ReSW reduced the average clearance time from 11 to 1.5 days. In 2020, the total saving for traders on direct cost to buy forms and pay clearing agents to manually fill the form and follow up the approval in the ministries exceeds 9 million USD.

Rwanda is landlocked, so the Rwanda Revenue Authority uses the ReSW to connect with the Port Authorities of Mombasa (Kenya) and Dar es Salaam (Tanzania) and has established offices in the East Africa Community Single Customs Territory. In addition, the ReSW is interlinked with the customs systems of Uganda and Kenya on the Northern Corridor and with the Tanzanian customs system on the Central Corridor. Once imports are processed, an exit note is issued through the single window and information is shared to the ports and the revenue authorities, enabling them to clear the goods. The ReSW relies on the corridor management institutions and also the Regional Electronic Cargo Tracking System which since 2020 has helped track and trace goods on the Northern Corridor to and from the Port of Mombasa.

Single windows can also be built for maritime systems. A maritime national single window (MNSW) can be used to harmonize and exchange data among the relevant port agencies, providing a single point of electronic document submission for port clearance. In Singapore, for example, the Government, in partnership with the IMO, has recently launched a Single Window for Facilitation of Trade that is aligned with the WTO TFA and the IMO FAL Convention recommendations on the electronic exchange of data (see section F of this chapter).

NTFCs can facilitate communication and coordination among the different stakeholders to create synergies and ultimately establish single points of access along the supply chain covering transport and trade procedures.

Other IT applications designed to undertake pre-arrival processing such as ASYHUB expedite customs clearance procedures, and minimize the time and cost of trade operations (section C of this chapter).

Table 6.1	Key performance indicators of the Kenya Trade Information Portal	
Kenya Trade Information Portal (52 trade procedures)		
• 44 of 52 procedures have been simplified	• 50 steps eliminated 1.1 on average	• 20 steps now accessible online • 66% of all steps are now online (baseline: 46%)
• 110 hours saved 2.5 on average	• 53,000 KES saved fees ($480 saved) 1,205 KES average reduction ($10.9)	• 66 documents eliminated 1.5 on average

Source: Kenya Trade Information Portal, https://infotradekenya.go.ke.

Another ICT innovation, based on UNCTAD technology, is the Trade Information Portal (TIP). Governments can use this online portal to document and publicize trade procedures for export, import and transit. Each TIP offers step-by-step guides to trade-related procedures. The TIP, which is coordinated by the Secretariat of the National Trade Facilitation Committee, simplifies and streamlines procedures while increasing transparency of trade information on export, import and transit requirements. In this way countries can fulfil their obligations in WTO TFA, article 1.2 on information availability through the internet.

Today, 29 TIPs, based on UNCTAD technology, are being implemented globally by UNCTAD and the International Trade Centre. Results have been very positive. TIPs are most advanced in East Africa, where in Kenya, for example, greater transparency and simplification of a total of 52 trade procedures so far have reduced the time spent waiting in the queue, at the counter and in between steps by 110 hours, and the administrative fees for these 52 procedures by $482, i.e., about $11 per trade procedure on average (table 6.1).

An essential element of measures to improve trade facilitation is digitalization, which is part of a paperless environment. All trade procedures can then be carried out online, reducing time and cost for the traders and increasing transparency and market access. These smart solutions also enable better public administration of trade and, by minimizing the use of paper and carbon-based activities, can reduce CO_2 emissions (Duval, 2021). However, these benefits will only be achieved through sustained intergovernmental and public-private sector cooperation at all levels (box 6.1).

Box 6.1	The Framework Agreement on Facilitation of Cross-Border Paperless Trade in Asia and the Pacific - Maritime implications

The Framework Agreement on Facilitation of Cross-border Paperless Trade in Asia and the Pacific (ESCAP, 2021) aims to accelerate digitalization of trade in support of sustainable development. After four years of negotiations, the Economic and Social Commission for Asia and the Pacific (ESCAP) adopted the treaty in May 2016 and opened it to all its 53 member States.

The Agreement entered into force on 20 February 2021, following accession or ratification of Azerbaijan, the Philippines, the Islamic Republic of Iran, Bangladesh, and China. Armenia and Cambodia have also signed the treaty. Several other ESCAP member States are in the process of accession, in time for the first meeting of the Paperless Trade Council. This body will oversee the implementation of the Agreement starting in March 2022.

Designed as an enabling rather than a prescriptive instrument, the Agreement is accessible to countries at all levels of development. It contains general principles and other provisions to facilitate pilot testing and implementation of paperless trade solutions suitable for each country, while promoting interoperability across systems and public-private sector collaboration within and across borders. The Agreement complements the WTO Trade Facilitation Agreement and supports its full digital implementation. Trade cost reductions expected from the full implementation of cross-border paperless trade are estimated at 10-30 per cent of existing transactions costs, depending on the current state of paperless trade development in the participating countries (ESCAP, 2017).

This agreement will boost the digitalization of maritime transport in Asia and the Pacific, which is home to nine of world's ten busiest ports and has the bulk of global maritime trade. It should also provide a strong political and institutional basis to improve the interconnectivity of maritime single windows and port community systems. It will also help digitalize maritime documents such as bills of lading, packing lists and manifests that are used in governmental trade compliance and in processes agreed between traders and transport and logistics service providers. As these documents are digitalized, they need to be shared and legally recognized across both in maritime single window/port community systems and trade single window systems, and can be shared across all paperless systems along international supply chains. Backed by this agreement, the Paperless Trade Council can engage relevant international organizations, private sector stakeholders and development partners to fill the capacity gaps and facilitate interoperable solutions.

Source: ESCAP.

F. FAL CONVENTION

The WTO TFA addresses issues in relation to the clearance of goods. The Convention on Facilitation of International Maritime Traffic (FAL Convention), on the other hand, which is managed by the IMO, focuses on the formalities and procedures for ships calling in ports, including those related to the arrival and departure of seafarers. Trade facilitation initiatives are likely to involve both agreements, so careful coordination and integration will be needed at the national level in order to ensure that regulations and procedures are aligned.

1. Main provisions of the Convention

The FAL Convention has both compulsory and recommended provisions. Contracting governments can thus comply to the extent they are able to. One of its most important measures concerns the number of documents that shore authorities can require, which it limits to 12. For the first seven of these, the IMO has developed standardized forms, widely known as FAL forms, which include General Declaration (FAL Form 1), and Cargo Declaration (FAL Form 2). Nevertheless authorities can also require other documentation pertaining, for example, to the ship's registration, measurement, safety, pollution prevention, or safe manning. The FAL Convention also contains provisions to prevent, report on, and resolve stowaway incidents, as well and standards and recommendations on treatment of stowaways while on board ships.

For the FAL Convention, significant efforts have been made to promote digitalization, with new provisions to allow for data to be submitted and shared electronically. Since 2019, public authorities in ports must set up the electronic exchange of information, and may only use paper forms in exceptional circumstances. To reduce duplication, the FAL Convention also recommends the single window approach, aligned with Article 10.4 of the TFA, whereby ship reporting parties can fulfil the requirements of the various authorities by providing information once to a single entry point.

In 2021, the FAL Committee approved amendments to the Convention that further promote digitalization. Once these are formally adopted, the FAL Convention will no longer refer to paper forms but to a list of data requirements. In addition, the single window will become mandatory. These amendments are expected to be adopted by the FAL Committee in 2022 and to enter into force in January 2024.

The FAL Committee also aims to improve the quality of data exchange between ships and ports. An important contribution to this is the IMO Compendium on Facilitation and Electronic Business which provides a common terminology so that shipping and ports use the same definitions and formats. The IMO Compendium can also be used by other IMO Committees when preparing their requirements on electronic reporting and information exchange.

> **Box 6.2   IMO Compendium on Facilitation and Electronic Business**
>
> The IMO Compendium on Facilitation and Electronic Business aims to harmonize the essential standards for ship clearance and to support electronic data exchange between ships and ports. It was developed by the IMO in partnership with ECE, WCO and ISO.
>
> The Compendium has two critical components: the IMO Data Set (IDS) and the IMO Reference Data Model (IRDM). The IDS provides unique identification, and a common definitions and representations/formats for all the data elements. The IRDM defines how the data elements relate to each other – reflecting the relationships between the different areas of information.
>
> Initially, the IMO Compendium was limited to the FAL Convention (i.e., FAL forms). This led to a partnership agreement between ECE, WCO and ISO to develop and maintain the IRDM. To ensure full interoperability between the most relevant standards, the data elements are mapped across the main models – UN/CEFACT, WCO Data Model and ISO. The data exchange syntax for electronic messages, is provided by the corresponding organizations.
>
> Since 2019, the scope of the IMO Compendium has been extended. It now covers other IMO instruments (e.g., MARPOL and SOLAS) and other data specifications related to the ship/shore interface. Since 2020, the IMO Compendium has included the Maritime Declaration of Health (MDH), a requirement of the International Health Regulations (IHR) under the purview of the WHO. The IMO Compendium also includes IMO data on stowaways as well as operational and real-time data to help optimize port calls and decarbonize shipping. More data sets are currently being prepared by the IMO Expert Group on Data Harmonization, a group of Member States and industry experts set up to maintain the IMO Compendium. Data sets related to shipping certificates, ship registry and company details, ballast water reporting, and the verified gross mass of containers are being considered for inclusion in 2022.
>
> *Source:* IMO.

In 2021, having learned from the COVID-19 pandemic, Member States are adding a new section addressing a public health emergencies of international concern (PHEIC) to the FAL Convention. To help sustain global supply chains during a PHEIC, contracting governments and their relevant public authorities must ensure that ships and ports remain fully operational. And they should designate port workers and crew members who are in their territory as key workers or equivalent, regardless of their nationalities or the flag of their ship. National authorities are also advised not to introduce obstacles to crew movements for repatriation, crew changes or travel. The new amendments to be adopted in 2022 also encourage governments to disseminate information about public health matters and the protection measures expected from ship operators.

2. FAL Convention requirements for maritime single windows and port community systems

When a ship calls at a port, the master or the shipping agent has to fulfil regulatory and port entry requirements – for purposes of safety, security, and environmental protection. This includes submitting information on the ship, and its voyage, cargo, crew, and passengers. This information is used for various clearance and port call processes – including pre-arrival, arrival, berthing, loading/unloading, embarkation/disembarkation, clearance, and departure/unberthing.

Since 2019, the IMO has required this information to be exchanged electronically. On the ship this could involve the master, ship agents, and shipping lines, while those involved ashore include maritime administrations, and the authorities concerned with customs, police/law enforcement, immigration, public health, port administration, and agriculture.

The IMO also recommends that data is submitted through a single window using software that distributes the information to relevant stakeholders according to the system rules and user agreements. The single window in port covers business-to-government and government-to-business exchanges.

In 2019, IMO produced guidelines for setting up a maritime single window (MSW) to help Member States and software developers, with examples of different approaches in existing systems. (FAL.5/Circ.42/Rev.1). Developing such systems is complex and involves multiple stakeholders based on an appropriate legal framework for data requirements and sharing.

Other forms of eSW include national single windows (NSW) or customs or trade single windows (TSW). Possible gateways into the various systems are port community systems (PCS). As defined by International Port Community Systems Association (IPCSA), a PCS is a neutral and open electronic platform enabling the intelligent and secure exchange of information between public and private stakeholders.

Since 2019, IMO has encouraged Member States that are more advanced in MSW implementation to exchange know-how and experiences with other Member States seeking assistance. Norway, for example, has made available the source code of a generic maritime single window system developed as part of a project with the IMO. Its design is of particular interest to SIDS and it has been implemented in Antigua and Barbuda. It is accessible at https://github.com/Fundator/IMO-Maritime-Single-Window.

In 2021, the IMO launched two technical cooperation initiatives. One aims to develop and implement a maritime single window in a medium-size port based on Singapore's experience – the Single Window for Facilitation of Trade (SWiFT). In April 2021, there was call for interest to identify the pilot country. The second project is the 'World Bank Group/IMO maritime single window for SIDS' which will provide Fiji with technical support to adopt and implement an MSW based on the source code from Norway, and the experience of Antigua and Barbuda.[19]

The amendments to the FAL Convention approved in 2021 will make the use of the single window mandatory. Public authorities must also try to ensure that the information is submitted electronically only once and re-used as much as possible.

During the COVID-19 pandemic, a group of global industry associations in consultative status with the IMO representing the maritime transportation and port sectors agreed on a joint statement calling for intergovernmental collaboration to accelerate the digitalization of maritime trade and logistics. The IMO supported the joint statement and has encouraged collaboration between maritime supply chain industry stakeholders and Member States and called for intergovernmental collaboration at local, national, and regional levels.[20]

---

[19] https://www.imo.org/en/MediaCentre/PressBriefings/Pages/07-IMO-maritime-data-solution-available-after-launch-in-Antigua-and-Barbuda-.aspx.

[20] https://wwwcdn.imo.org/localresources/en/MediaCentre/HotTopics/Documents/COVID%20CL%204204%20adds/Circular%20Letter%20No.4204-Add.20%20-%20Coronavirus%20(Covid-19)%20-%20Accelerating%20Digitalization%20Of%20Maritime%20Trade.pdf.

G. ASYCUDA ASYHUB CASE STUDIES

The WTO TFA and the FAL Convention recognize the importance of automating and digitalizing customs and trade procedures – by focusing on issues such as eSW, port community systems, and overall interconnectivity and interoperability at national levels and across borders. This section provides examples of the practical implementation of these aspects based on experience from UNCTAD's ASYCUDA.

ASYCUDA is a computerized customs management system that covers most foreign trade procedures. It handles manifests and customs declarations, accounting procedures, and transit and suspense procedures. It also generates trade data that can be used for statistical analysis.

Many customs administrations have introduced procedures for submitting cargo information in advance, in line with the obligations of the WTO TFA. However, this is typically submitted only 24 hours before arrival, leaving customs administrations little time for risk assessment and processing – and potentially increasing turnaround times for traders, logistics operators and freight forwarders.

The information pertaining to a shipment is logged many weeks in advance but this data may not be accessible to all the organizational entities needed to grant customs clearances. ASYCUDA facilitates the sharing of this information in advance to enable customs to clear goods upon arrival, generally plan better, and reduce overall clearance times.

1. Digitizing Global Maritime Trade

To enhance further risk-based pre-arrival/pre-departure processing, the Digitizing Global Maritime Trade (DGMT) project[21] focuses on enabling customs authorities to gain advance digital access to sea cargo information (PAP/PDP) – as stipulated in WTO TFA Articles 7.1 and 7.4.

Started by UNCTAD/ASYCUDA in December 2019 in partnership with Deutsche Gesellschaft für Internationale Zusammenarbeit (GIZ) and the shipping industry in the context of the German Trade Alliance for Trade Facilitation, the DGMT project aims at:

- Increasing efficiency in the international transport documentation process
- Reducing the time and costs of maritime trade for importers and exporters
- Streamlining risk management by increasing digital access by customs authorities to advance sea cargo information during clearance processes

---

Box 6.3   Components of the Digitizing Global Maritime Trade project

1. Development of ASYHUB Maritime, a standardized data exchange and data integration platform between ASYCUDAWorld and international standards-compliant shipping data platforms. The objective of component 1 is to harmonize and streamline information exchange between international standards-compliant data platforms and customs administrations. This allows for the efficient transfer of advanced cargo information and for existing data to be reused to complete the entry/exit customs formalities. The ASYHUB Maritime platform is now ready for piloting.

2. Enhance the capacity of customs authorities in Sri Lanka and Cambodia to apply ASYHUB Maritime to improve pre-arrival and pre-departure processing and risk management. This component aims to improve their risk management systems by using new datasets and new technology solutions. Customs authorities can then conduct risk assessments and process cargo and customs declarations prior to the arrival of goods at the port of entry/port of exit. This will enable the release of the cleared goods shortly after arrival.

3. Outreach to create demand and initiate upscaling to at least five further countries during or shortly after the successful conclusion of the first two pilots.

The two pilot countries will share their experiences with the network and receive advice and expertise from their peers. The five early adopter countries can take steps towards pre-arrival and pre-departure processing and risk management through ASYHUB Maritime and international standards-compliant shipping data providers.

Source: UNCTAD ASYCUDA.

---

[21] Grant Agreement #81249048 between GIZ and UNCTAD/ASYCUDA signed in October 2019.

This project involves the development of ASYHUB Maritime, a standardized data exchange and integration platform. Currently, the project is in phase two of a three-phase process and is being testing in two pilot countries. This will be followed by the creation of a virtual community of practice consisting of countries using ASYCUDA World, to enable its potential replication or upscaling in over 90 countries.

2. ASYHUB and single window integration

ASYHUB Maritime is an open, standardized platform for data processing and data integration between ASYCUDAWorld and other external systems. The platform is designed to be cloud-native using micro service-centred principles. It will simplify and automate the submission of sea cargo manifest information through a system-to-system interface, providing customs authorities with richer information that can be used to make informed risk assessments and better decisions on which shipments to inspect. This will reduce the administrative burden for ship data providers, increase trade facilitation, ensure a quicker release of goods, and improve risk management, security, and revenue collection.

The ASYHUB Maritime platform enables ship data providers to re-use the existing data to complete the entry/exit formalities and exchange advanced electronic cargo information with port authorities, customs, and other border agencies (box 6.4). This will also ensure better interconnectivity and interoperability between countries.

Box 6.4 Customs formalities concerning entry or exit

- Entry of goods
- Customs Cargo Manifest (at arrival)
- Arrival notification
- Presentation notification
- Temporary Storage Declaration
- Exit of goods
- Customs Cargo Manifest (at departure)
- Exit notification.

Source: UNCTAD ASYCUDA.

H. SUMMARY AND POLICY CONSIDERATIONS

Ensuring maritime cybersecurity

The maritime sector is increasingly structured around online and automated systems. Recently updated industry guidelines offer shipowners and operators information on procedures and actions to maintain cybersecurity in their companies and ships – adopting cyber-risk management approaches that take account of IMO requirements and other relevant guidelines. Implementing cybersecurity not only helps shipowners avoid having their ships detained by port State control authorities, it also makes economic sense, and helps protect shipping assets and technology from increasing cyber-threats.

Regulating maritime autonomous surface ships

The industry is advancing rapidly with the technology for maritime autonomous surface ships (MASS) and is now conducting trials. In May 2021, the IMO Maritime Safety Committee completed a regulatory scoping exercise which highlighted high-priority issues that cut across several instruments and will need policy decisions to determine future work. This could result in a MASS instrument or code, with goals, functional requirements and corresponding regulations. Developing countries representatives and other stakeholders are encouraged to contribute to future discussions.

Reducing greenhouse gas emissions and adapting to climate change

Mitigation and adaptation to global climate change are increasingly urgent imperatives. Resilience-building is especially important for seaports that are exposed to sea-level rise and related extreme weather events. The 2021 IPCC report warns of increasingly extreme heatwaves, droughts, and flooding. Nevertheless, rising temperatures could be stabilized by deep cuts in emissions of GHGs in which shipping must play its part. In June 2021, the IMO adopted mandatory regulations that aim to cut the carbon intensity of ships and their carbon emissions. These include requirements to measure the energy efficiency of all ships and set the required attainment values. Adaptation remains a particular concern for vulnerable developing countries, including SIDS.

Reducing pollution from shipping

In 2020 the IMO set a 0.5 per cent sulphur limit on ship fuel oils. Flag and Port State controls need to make sure ships are compliant. During 2020 and the first half of 2021, implementation was relatively smooth with VLSFO as the preferred solution, and compliant fuel oil was widely available globally. Another major fuel oil concern is the risk of oil spills which can have devastating consequences for ecosystems and biodiversity and for the economies and tourist industries of coastal countries, which should be able to claim adequate compensation. Unfortunately, the very comprehensive international regime on liability and compensation for tanker oil spills (CLC-IOPC Fund regime), does not apply to bunker oil spills from other types of ship. Given the continuing growth in the size of vessels of any type and the associated potential for significant bunker oil pollution, with devastating consequences for vulnerable coastal developing countries and SIDS, the issue of liability for bunker oil spills from ships other than tankers may need to be revisited. The IMO is developing a claims manual for the Bunker Oil Pollution Convention, 2001 which addresses liability for bunker oil spills.

Commercial law implications of the pandemic, and the use of electronic trade documents

The COVID-19 pandemic continues to interfere with international trade, creating inefficiencies, delays and supply-chain disruptions on an unprecedented scale. This also has implications for contractual performance with potential legal consequences and litigation involving complex international jurisdictional issues. Resolving these problems will require collective and coordinated action by governments and industry. This could involve, for example agreeing contract extensions, showing restraint in pursuing rights and legal claims, and resolving disputes through mediation and informal mechanisms. It could also involve commercial risk allocation through standard clauses to address contractual rights and obligations in the light of the circumstances associated with the pandemic. Recent UNCTAD reports provide analytical guidance to commercial parties and governments on some of the key legal issues arising.

Digitalizing trade facilitation

Maritime transport can be impeded by regulatory requirements and slow clearance procedures at ports. Trade facilitation can, however, be improved by digitalization and automation of customs and other compliance processes, single window implementation, ensuring that formalities are increasingly paperless. Frameworks and common standards and regulations for these systems can be based on multilateral agreements, e.g., through the WTO TFA and the IMO FAL Convention.

Connectivity requires cooperation and coordination

New technologies and smart solutions raise questions of interconnectivity and interoperability and the need for international standards. When digitalizing and automating their systems, developing and least developed countries can take advantage of the experiences of other countries and follow good practices already available, such as those of the ASYCUDA system.

National trade facilitation committees

Any successful trade reform relies on cooperation between public administrations and the business community. For this purpose, each country should set up an NTFC comprising public and private stakeholders at national levels who should devise a coherent and coordinated strategy and champion and drive the trade facilitation agenda. The NTFC membership should represent all the businesses involved in maritime trade and port operations who can work with the government authorities to make logistics supply chains more efficient and boost national trade performance.

REFERENCES

Asariotis R, Premti A (2020). Mauritius oil spill highlights importance of adopting latest international legal instruments in the field. 14 August. Available at https://unctad.org/news/mauritius-oil-spill-highlights-importance-adopting-latest-international-legal-instruments.

BIMCO (2018). 2020 Marine Fuel Sulphur Content Clause for Time Charter Parties. 10 December. Available at https://www.bimco.org/contracts-and-clauses/bimco-clauses/current/2020_marine_fuel_sulphur_content_clause_for_time_charter_parties.

BIMCO et al (2018). The Guidelines on Cyber Security Onboard Ships. Version 3. Available at http://www.ics-shipping.org/docs/default-source/resources/safety-security-and-operations/guidelines-on-cyber-security-onboard-ships.pdf?sfvrsn=16.

BIMCO (2021). BIMCO, NGOs invite the industry to help develop global digital ISO standards. 5 March. Available at https://www.bimco.org/news/maritime-digitalisation/20210305-global-digital-iso-standards.

BIMCO et al (2021). The Guidelines on Cyber Security Onboard Ships. Version 4. Available at https://www.bimco.org/about-us-and-our-members/publications/the-guidelines-on-cyber-security-onboard-ships.

British Ports Association (2020). Managing ports' cyber risks. White Paper. Available at https://astaara.co.uk/wp-content/uploads/2020/07/Astaara-White-Paper-v8.pdf.

Dan Rutherford, X M (2020). Limiting engine power to reduce CO2 emissions from existing ships. ICCT.

DCSA (2020). DCSA Implementation Guide for Cyber Security on Vessels v1.0. 10 March. Available at https://dcsa.org/wp-content/uploads/2020/03/DCSA-Implementation-Guideline-for-BIMCO-Compliant-Cyber-Security-on-Vessels-v1.0.pdf.

Dibke C, Fischer M, Scholz-Böttcher B M (2021). Microplastic Mass Concentrations and Distribution in German Bight Waters by Pyrolysis–Gas Chromatography–Mass Spectrometry/Thermochemolysis Reveal Potential Impact of Marine Coatings: Do Ships Leave Skid Marks? Environmental Science and Technology, 55 (4): 2285–2295.

Duval Y (2021). A primer on quantifying the environmental benefits of cross-border paperless trade facilitation. Bangkok: ARTNeT Working Paper Series No. 206.

ESCAP (2017). *Digital Trade Facilitation in Asia and the Pacific*. Bangkok: United Nations. 22 December. Retrieved from https://www.unescap.org/publications/digital-trade-facilitation-asia-and-pacific-studies-trade-investment-and-innovation-87#.

ESCAP (2020). Virtual Expert Group Meeting on Legal Frameworks for Multimodal Transport Operations in Asia and the Pacific. 26–27 August. Available at https://unescap.org/events/virtual-expert-group-meeting-legal-frameworks-multimodal-transport-operations-asia-and.

ESCAP (2021). Virtual Expert Group Meeting on Legal Frameworks for Multimodal Transport Operations in Asia and the Pacific. 30–31 March. Available at https://www.unescap.org/events/2021/second-virtual-expert-meeting-legal-frameworks-multimodal-transport-operations-asia-and.

ESCAP (2021, June 30). *Framework Agreement on Facilitation of Cross-border Paperless Trade in Asia and the Pacific*. Retrieved from https://www.unescap.org/resources/framework-agreement-facilitation-cross-border-paperless-trade-asia-and-pacific.

European Commission, 2021. Communication from the Commission to the European Parliament, the Council, the European Economic and Social Committee and the Committee of the Regions - Forging a climate-resilient Europe - the new EU Strategy on Adaptation to Climate Change. Available at https://eur-lex.europa.eu/legal-content/EN/TXT/HTML/?uri=CELEX:52021DC0082&from=EN.

European Union (2021). Regulation (EU) 2021/1119 of the European Parliament and of the Council of 30 June 2021 establishing the framework for achieving climate neutrality and amending Regulations (EC) No 401/2009 and (EU) 2018/1999 ('European Climate Law'). Available at https://eur-lex.europa.eu/legal-content/EN/TXT/PDF/?uri=CELEX:32021R1119&from=EN.

Gard (2020). Another step towards a zero-emission future. 21 December. Available at https://www.gard.no/web/updates/content/30906149/another-step-toward-a-zero-emission-future.

Hallegatte S, Rentschler J, Rozenberg J (2019). Lifelines: The Resilient Infrastructure Opportunity. Sustainable Infrastructure;. Washington, DC: World Bank. © World Bank. Available at https://openknowledge.worldbank.org/handle/10986/31805.

IACS (2020). Recommendation on Cyber Resilience (No. 166). Available at https://www.google.com/url?sa=t&rct=j&q=&esrc=s&source=web&cd=&ved=2ahUKEwiKl56C4tbwAhUh7eAKHbuqDF4QFjACegQIAxAD&url=https%3A%2F%2Fwww.iacs.org.uk%2Fdownload%2F10965&usg=AOvVaw3BXnWt78eg2uQPPY0ZNY3G.

IMO (2017). Report of the Maritime Safety Committee on its ninety-eighth session. MSC 98/23. London. 28 June.

IMO (2018). Report of the Working Group on Reduction of greenhouse gas emissions from ships. MEPC 72/WP.7. London.

IMO (2019a). Proposal to add a new output under the work programme on "Unified Interpretation on the test for breaking the owner's right to limit liability under the IMO conventions". LEG 106/13. London. 11 January.

IMO (2019b). Report of the Legal Committee on the work of its 106th session. LEG 106/16. London. 13 May.

IMO (2020a). Reduction of GHG emissions from ships. Fourth IMO GHG Study 2020 – Final report. MEPC 75/7/15. 29 July. London.

IMO (2020b). Report of the Legal Committee on the work of its 107th session. LEG 107/18/2. 11 December.

IMO (2021a). Guidelines on maritime cyber risk management. MSC-FAL.1/Circ.3/Rev 1. 14 June. Available at https://wwwcdn.imo.org/localresources/en/OurWork/Facilitation/Facilitation/MSC-FAL.1-Circ.3-Rev.1.pdf.

IMO (2021b). Report of the Maritime Safety Committee on its 103rd session. MSC 103/21. London. 25 May.

IMO (2021c). Report of the Legal Committee on the work of its 108th session. LEG 108/16/1. 25 August.

IMO (2021d). Report of the Marine Environmental Protection Committee on its 76th session. MEPC 76/15. London. 12 July.

IMO (2021e). Reduced limit on sulphur in marine fuel oil implemented smoothly through 2020. 28 January. Available at https://www.imo.org/en/MediaCentre/PressBriefings/Pages/02-IMO-2020.aspx.

InfoPort (2021, June 30). Retrieved from https://www.infoport.es/port-community-system-en/?lang=en.

IOPC FUNDS (2019). Claims Manual. October. Available at https://iopcfunds.org/wp-content/uploads/2018/12/2019-Claims-Manual_e-1.pdf.

INTERTANKO (2018). Bunker Compliance Clause for Time Charterparties. 7 December. Available at https://www.intertanko.com/info-centre/model-clauses-library/templateclausearticle/intertanko-bunker-compliance-clause-for-time-charterparties.

IPCC (2018). Special Report on Impacts of 1.5 °C global warming. Available at https://www.ipcc.ch/sr15/.

IPCC (2019). Special Report on Ocean and Cryosphere. Available at https://www.ipcc.ch/srocc/download/.

IPCC (2021). Climate Change 2021. The Physical Science Basis. Available at https://www.ipcc.ch/report/ar6/wg1/downloads/report/IPCC_AR6_WGI_Full_Report.pdf.

ISO (2021). ISO 14091:2021 - Adaptation to climate change-Guidelines on vulnerability, impacts and risk assessment. Available at https://www.iso.org/standard/68508.html.

NIST (2018). Framework for Improving Critical Infrastructure Cybersecurity. Version 1.1. 16 April. Available at https://www.nist.gov/cyberframework/framework.

OECD (2017). Investing in Climate, Investing in Growth. Available at https://www.oecd.org/environment/investing-in-climate-investing-in-growth-9789264273528-en.htm.

OECD/WTO. (2019). Aid for Trade at a Glance 2019: Economic Diversification and Empowerment. Paris: OECD Publishing.

PIANC (2019). PIANC Declaration on Climate Change. Available at https://www.pianc.org/uploads/files/COP/PIANC-Declaration-on-Climate-Change.pdf.

PIANC (2020). PIANC [World Association for Waterborne Transport Infrastructure] Climate change adaptation planning for ports and inland waterways. Available at https://www.pianc.org/publications/envicom/wg178.

Ugaz P (2019, February 22). The World Trade Organization's Trade Facilitation Agreement at two: Where do members stand? Geneva: UNCTAD.

UNCITRAL (2019). Report of UNCITRAL on the work of its 52nd session. A/74/17. Available at https://undocs.org/en/A/74/17.

UNCITRAL (2020a). Possible future work on railway consignment notes. Note by the Secretariat. 11 May. Available at https://undocs.org/en/A/CN.9/1034.

UNCITRAL (2020b). Report of UNCITRAL on the work of its 53rd session. A/75/17. Available at https://undocs.org/en/A/75/17.

UNCITRAL (2021a). Results of the preparatory work by the UNCITRAL secretariat towards the development of a new international instrument on negotiable multimodal transport documents. Note by the Secretariat. A/CN.9/1061. 4 May. Available at https://undocs.org/en/A/CN.9/1061.

UNCITRAL (2021b). Report of UNCITRAL on the work of its 54th session. A/76/17. Available at https://undocs.org/en/A/76/17.

UNCTAD (2003). Multimodal Transport: the Feasibility of an International Legal Instrument. UNCTAD/SDTE/TLB/2003/1 13 January. Available at https://unctad.org/system/files/official-document/sdtetlb20031_en.pdf.

UNCTAD (2011). The 2004 Ballast Water Management Convention – with international acceptance growing, the Convention may soon enter into force. In: Transport Newsletter No. 50.

UNCTAD (2012). Liability and Compensation for Ship-Source Oil Pollution: An Overview of the International Legal Framework for Oil Pollution Damage from Tankers. Available at https://unctad.org/system/files/official-document/dtltlb20114_en.pdf.

UNCTAD (2015a). *Review of Maritime Transport 2015*. UNCTAD/RMT/2015. United Nations publication. Sales no. E. 15.II.D.6.

UNCTAD (2015b). The International Ballast Water Management Convention 2004 is set to enter into force in 2016. In: Transport and Trade Facilitation Newsletter No. 68.

UNCTAD. (2016). Trade facilitation and development: Driving trade competitiveness, border agency effectiveness and strengthened governance. Geneva.

UNCTAD (2018). *Review of Maritime Transport 2018* (United Nations publication. Sales No. E.18.II.D.5. New York and Geneva.

UNCTAD (2019). *Review of Maritime Transport 2019* (United Nations publication. Sales No. E.19.II.D.20. New York and Geneva.

UNCTAD (2020a). *Review of Maritime Transport 2020*. UNCTAD/RMT/2020. Available at https://unctad.org/system/files/official-document/rmt2020_en.pdf.

UNCTAD (2020b). Mauritius oil spill puts spotlight on ship pollution. 19 August. Available at https://unctad.org/news/mauritius-oil-spill-puts-spotlight-on-ship-pollution.

UNCTAD (2020c). Multi-year expert meeting on transport, trade logistics and trade facilitation, eighth session, 27–28 October. Available at https://unctad.org/meeting/multi-year-expert-meeting-transport-trade-logistics-and-trade-facilitation-eighth-session.

UNCTAD (2020d). Report of the Multi-year Expert Meeting on Transport, Trade Logistics and Trade Facilitation on its eighth session. Available at https://unctad.org/system/files/official-document/cimem7d24_en.pdf.

UNCTAD (2021a). UNCTAD assessment of the impact of the IMO Short-Term Greenhouse Gas Reduction Measure on States: Assessment of impacts on maritime logistics cost, trade and GDP. UNCTAD/DTL/TLB/2021/2. Available at https://unctad.org/system/files/official-document/dtltlb2021d2_en.pdf.

UNCTAD (2021b). UNCTAD15 pre-event: Harnessing the benefits of the ocean economy for sustainable development. 9 June. Available at https://unctad.org/meeting/unctad15-pre-event-harnessing-benefits-ocean-economy-sustainable-development.

UNCTAD (2021c). Leading the push for a sustainable ocean economy. 22 June. Available at https://unctad.org/news/leading-push-sustainable-ocean-economy.

UNCTAD (2021d). COVID-19 implications for commercial contracts: Carriage of goods by sea and related cargo claims. UNCTAD/DTL/TLB/INF/2021/1. 2 March. Available at https://unctad.org/webflyer/covid-19-implications-commercial-contracts-carriage-goods-sea-and-related-cargo-claims.

UNCTAD (2021e). COVID-19 implications for commercial contracts: International sale of goods on CIF and FOB terms. UNCTAD/DTL/TLB/INF/2021/2. 2 March. Available at https://unctad.org/webflyer/covid-19-implications-commercial-contracts-international-sale-goods-cif-and-fob-terms.

UNCTAD Business Facilitation Program. (2021, June 30). Available at https://unctad.org/topic/enterprise-development/business-facilitation.

UNFCCC (2021a). Climate action pathway. Transport. Action table. Available at https://unfccc.int/sites/default/files/resource/Transport_ActionTable_2.1.pdf.

UNFCCC (2021b). Climate action pathway: Transport. Vision and Summary. Available at https://unfccc.int/sites/default/files/resource/Transport_Vision%26Summary_2.1..pdf.

WTO (2021, June 30). Retrieved from WTO Trade Facilitation Agreement Database: https://tfadatabase.org/implementation/progress-by-measure.

Yara (2020). Yara Birkeland press kit. November. Available at https://www.yara.com/news-and-media/press-kits/yara-birkeland-press-kit/.